MALL ENGINES
operation and service

Jay Webster

AMERICAN TECHNICAL PUBLISHERS, INC.
HOMEWOOD, ILLINOIS 60430

Library of Congress Catalog Number: 80-67346
ISBN 0-8269-0004-6

123456789-81-9

PRINTED IN THE UNITED STATES OF AMERICA

contents

note to the student

SMALL ENGINES: OPERATION AND SERVICE is designed to help you learn about small engines. The purpose of this book is to help you train for an entry-level vocational apprenticeship, retrain or upgrade your skills or just help you learn about your own small engine.

This book is divided into eight parts. In PART 1, INTRODUCTION, we will begin by looking at how to work safely in the shop. You will learn how to use tools, fasteners and measuring systems important in small-engine service work.

IN PART 2 you will find out how small engines work. There are four-stroke-cycle, two-stroke cycle, diesel and rotary engines.

The systems of the small engine are presented in PART 3. You will learn how the ignition, fuel, lubrication and cooling systems of a small engine work.

PART 4 presents the principles of small engine size and performance measurement.

Starting in PART 5 you will begin the hands-on part of small engines. You will learn how to troubleshoot small engine problems, how to do periodic maintenance as well as how to tune-up the ignition and fuel systems.

PART 6 covers engine service. Here you will learn how to disassemble, service and reassemble a small engine.

In PART 7 you will learn about more specialized engines like motorcycles, mopeds, snowmobiles and outboards.

A career in small engines is the subject of PART 8.

The organization of each chapter is designed to make learning about small engines as easy as possible. Each chapter starts out with LET'S FIND OUT. Here you will find out what you should know or be able to do when you have read the chapter. At the end of each chapter there is a section of NEW TERMS. Each new term introduced in the chapter is defined again here to help you learn the terms used in small engine work.

Each chapter has a section called SELF CHECK. This section has questions on the main points of the chapter. SELF CHECK will help you find out if you have understood the main points of the chapter.

Finally, each chapter has some DISCUSSION TOPICS AND ACTIVITIES. You may find here some activities that interest you and expand your knowledge of small engines.

how to read service manuals

When you repair a small engine, you will find that many of the parts require careful assembly and adjustment. The repair of these engines would be impossible without up-to-date service manuals. Service manuals are books that list step-by-step repair procedures and specifications. Specifications, sometimes called "specs," are measurements and dimensions the small-engine manufacturer recommends for various parts. You must know how to find specs when you need them.

Here are the manuals you will be using:

Owner's Manual — This is a pamphlet that comes with a new engine. It gives instructions for use and lists some basic services and specs for the engine.

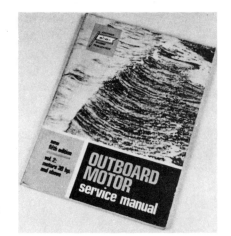

General Repair Manual — This manual has specs and step-by-step repair instructions for many different brands of small engines.

Manufacturer's Shop Manual — This comes from the manufacturer of the engine. It has the most detailed step-by-step repair instructions and the most up-to-date specs.

identifying your engine

When you want to look up specifications for your engine, you must first find out exactly what engine you have.

First look for the *brand name* of the manufacturer. This often is found on a large decal on the blower housing.

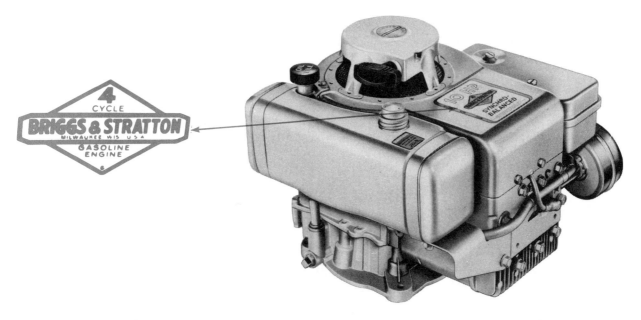

Secondly, look for the engine *model number*. The model number often is stamped on the blower housing. On some engines, it is stamped on a metal tag attached to the crankcase. Both serial and part numbers are on the engine. Make sure your number is the model number.

MODEL NUMBER

USING YOUR MODEL NUMBER

The model number not only helps you find specs for your engine; it also tells something about the engine. Each engine manufacturer uses a different model number code. You can find it in the manufacturer's shop manual. The model code usually tells the engine size and something about its equipment. Here is the code system for Briggs & Stratton:

A. The first one or two digits indicate the CUBIC INCH DISPLACEMENT.
B. The first digit after the displacement indicates BASIC DESIGN SERIES, relating to cylinder construction, ignition, general configuration, etc.
C. The second digit after the displacement indicates POSITION OF CRANKSHAFT AND TYPE OF CARBURETOR.
D. The third digit after the displacement indicates TYPE OF BEARINGS and whether or not the engine is equipped with REDUCTION GEAR or AUXILIARY DRIVE.
E. The last digit indicates the TYPE OF STARTER

CUBIC INCH DISPLACEMENT	FIRST DIGIT AFTER DISPLACEMENT — BASIC DESIGN SERIES	SECOND DIGIT AFTER DISPLACEMENT — CRANKSHAFT, CARBURETOR GOVERNOR	THIRD DIGIT AFTER DISPLACEMENT — BEARINGS, REDUCTION GEARS & AUXILIARY DRIVES	FOURTH DIGIT AFTER DISPLACEMENT — TYPE OF STARTER
6	0	0 -	0 - Plain Bearing	0 - Without Starter
8	1	1 - Horizontal Vacu-Jet	1 - Flange Mounting Plain Bearing	1 - Rope Starter
9	2	2 - Horizontal Pulsa-Jet	2 - Ball Bearing	2 - Rewind Starter
10	3			
11	4	3 - Horizontal Flo-Jet (Pneumatic Governor)	3 - Flange Mounting Ball Bearing	3 - Electric - 110 Volt, Gear Drive
13	5			
14	6	4 - Horizontal Flo-Jet (Mechanical Governor)	4 -	4 - Elec. Starter-Generator - 12 Volt, Belt Drive
17	7			
19	8	5 - Vertical Vacu-Jet	5 - Gear Reduction (6 to 1)	5 - Electric Starter Only - 12 Volt, Gear Drive
20	9			
23		6 -	6 - Gear Reduction (6 to 1) Reverse Rotation	6 - Alternator Only *
24				
25		7 - Vertical Flo-Jet	7 -	7 - Electric Starter, 12 Volt Gear Drive, with Alternator
30				
32		8 -	8 - Auxiliary Drive Perpendicular to Crankshaft	8 - Vertical-pull Starter
		9 - Vertical Pulsa-Jet	9 - Auxiliary Drive Parallel to Crankshaft	* Digit 6 formerly used for "Wind-Up" Starter on 60000, 80000 and 92000 Series

EXAMPLES

To identify Model 100202:

10	0	2	0	2
10 Cubic Inch	Design Series 0	Horizontal Shaft - Pulsa-Jet Carburetor	Plain Bearing	Rewind Starter

Similarly, a Model 92998 is described as follows:

9	2	9	9	8
9 Cubic Inch	Design Series 2	Vertical Shaft - Pulsa-Jet Carburetor	Auxiliary Drive Parallel to Crankshaft	Vertical Pull Starter

USING YOUR MODEL NUMBER TO FIND SPECIFICATIONS

When you know the brand name and model number for an engine, you can look up specifications in a General Repair Manual or a Manufacturer's Shop Manual. Turn to the page in the manual that lists specifications for your brand of engine. Let's say you have a Briggs & Stratton engine with model number 92000 and you want to know to what torque a cylinder head should be tightened:

| | BASIC MODEL SERIES | IDLE SPEED | ARMATURE | | VALVE CLEARANCE | | VALVE GUIDE REJECT GAGE | TORQUE SPECIFICATIONS | | |
			TWO LEG AIR GAP	THREE LEG AIR GAP	INTAKE	EXHAUST		FLYWHEEL NUT FT. LBS.	CYLINDER HEAD IN. LBS.	CONN. ROD IN. LBS.
A	6B, 60000	1750	.006 .010	.012 .016	.005 .007	.009 .011	19122	55	140	100
L	8B, 80000, 82000	1750	.006 .010	.012 .016	.005 .007	.009 .011	19122	55	140	100
U	92000	1750	.006 .010		.005 .007	.009 .011	19122	55	140	100
M	100000	1750	.010 .014	.012 .016	.005 .007	.009 .011	19122	60	140	100
I	110000	1750	.006 .010		.005 .007	.009 .011	19122	55	140	100
N	130000	1750	.010 .014		.005 .007	.009 .011	19122	60	140	100
U	140000	1750	.010 .014	.016 .019	.005 .007	.009 .011	19151	65	165	165
M	170000, 171700●	1750 **	.010 .014		.005 .007	.009 .011	19151	65	165	165
	190000, 191700●	1750 **	.010 .014		.005 .007	.009 .011	19151	65	165	165
	251000	1750	.010 .014		.005 .007	.009 .011	19151	65	165	190

First read down this model column to find your model number.

Then go over to the column for cylinder head torque specifications

Reading over from model number 92000 and down from cylinder head torque your specification is 140 inch-pounds

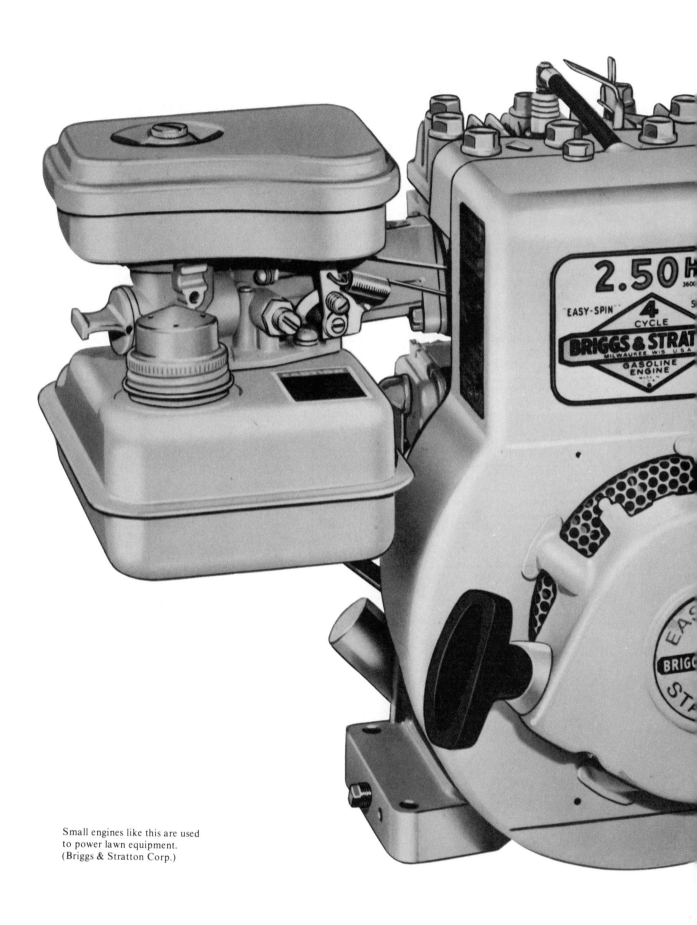

Small engines like this are used
to power lawn equipment.
(Briggs & Stratton Corp.)

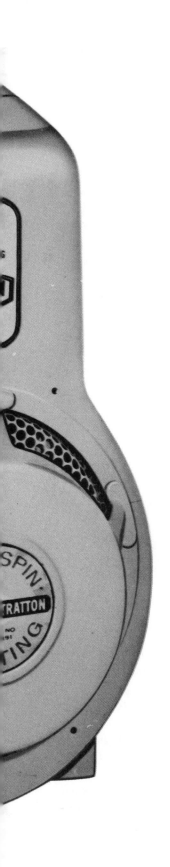

part 1
introduction — what is a small engine?

At first glance a small engine looks like a complicated piece of machinery. It is constructed of many parts and components and many different systems. However, each of the components and systems has purpose and operates on some very basic rules or principles.

An engine is a machine that changes burning fuel into power. In an engine, a fuel such as gasoline is burned to develop heat. The heat is then used to develop power. Power is used to do work.

With an engine you can:
- mow a lawn
- drive a car
- power a motorcycle
- ride a mini-bike
- fly in a helicopter
- ski behind a boat
- ride through the snow.

Engines used to power lawn or garden equipment are considered small engines. Engines used in small vehicles such as mopeds, motorcycles and snowmobiles are also considered small engines. Most small engines work in the same way. All have about the same parts. If you learn how a small lawnmower engine works, you will know how a car or motorcycle engine works. When you find out how a small engine gets fuel you will know how a car or motorcycle engine gets fuel.

The uses for small engines are many and varied. They are used to do work on equipment such as lawnmowers, edgers, generators, water pumps and chain saws. Small engines are used to power many types of recreational vehicles such as snowmobiles, snow trikes, trail bikes, mini bikes, home-built aircraft and outboards. Small engines are becoming more and more popular for transportation vehicles on mopeds and motorcycles. The things you will learn in this book will help you understand each of these power machines.

YOU are the key to SAFETY!

unit 1
working safely

When you work on small engines you will be working in a shop or lab. You will work near many things that must be used carefully. Working in the shop is not dangerous as long as you always keep in mind *safety first*. Before you do any job, always ask yourself if what you are doing is safe. If you do not know, ask your teacher. In this unit you will find out how to prevent accidents.

LET'S FIND OUT: When you finish reading and studying this unit you should be able to:
1. Demonstrate your knowledge of safety practices by working safely.
2. Recognize the common hazards in using hand and power tools.
3. Explain how to use cleaning equipment safely.
4. Describe the hazards involved in running an engine in the shop.
5. Explain how fires are prevented and extinguished.

USING HAND TOOLS

Believe it or not, many accidents are caused by the improper use of ordinary hand tools. Greasy tools can slip out of the hand. They may fall into a moving part of an engine and fly out to injure someone. Some tools have sharp edges that may cut you if you do not use them correctly. Follow these simple rules when using hand tools:
1. Be sure your hands are free of dirt, grease and oil.
2. Use the proper type and size hand tool.
3. Make sure the tools are sharp and in good condition.
4. Use sharp-edged or pointed tools with special care.
5. Make sure to point the edge of a sharp or pointed tool away from yourself and others.
6. Wear eye protection when filing or cutting metal, Figure 1-1. Arrange your work so that other people are protected from flying chips.

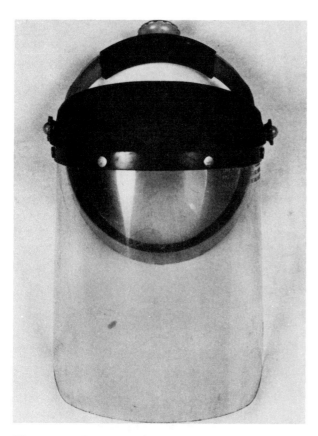

Figure 1-1. Eye protection should be worn when there is danger of flying particles.

7. Pass tools to others with the handles toward them.
8. Clamp small workpieces on a bench or secure them in a vise when driving screws.

USING POWER TOOLS

Tools which are powered by electricity, hydraulic fluid or compressed air are called power tools. Power tools move very fast, and if not used properly, may cause very serious injuries. The following rules apply to all power tools:

1. Obtain permission from your instructor before using any power tool.
2. Check adjustments on machines before turning on the power. Whenever possible without causing danger, rotate the machine one revolution by hand.
3. Make sure that no one is near the tool when you turn on the power.
4. Always wear eye protection when operating a power tool.
5. Keep all machine safety guards in correct position.
6. Start your own power tool and remain with it until you have turned it off and it has come to a complete stop.
7. Stay clear of power tools being operated by others.
8. Notify the instructor when a machine does not seem to work properly.
9. Wait for power tools to come to a complete stop before oiling, cleaning or adjusting. If possible, unplug the tool first.
10. Be sure clothes are safe and suitable for shop work. Remove or fasten any loose clothing or jewelry. Roll loose sleeves above the elbows. Keep hair away from equipment in operation.
11. Observe rules concerning safety lines around equipment.

USING CLEANING EQUIPMENT

Almost every repair operation requires that parts be cleaned. Most shops have several methods to clean parts. You may have to use a steam cleaner, cold tank, hot tank or solvent tank.

The main danger from cleaning equipment is getting harmful chemicals on the skin or in the eyes. Burns are an added danger when using the hot tank or steam cleaner. To avoid injury, follow these rules:

1. Obtain permission from your instructor before using any cleaning equipment.
2. Never use gasoline to clean parts.
3. When operating the steam cleaner, use a shop apron, face shield and gloves to protect from burns and splash.
4. Wear a shop apron, rubber gloves and a face shield when taking parts in or out of a hot or cold cleaning tank.
5. Wear protective clothing when cleaning parts to prevent chemical splash on the skin.

RUNNING AN ENGINE IN THE SHOP

You will be working on engines in the shop. Sometimes the engines will be running. If you run an engine at too fast a speed it could explode. Never run an engine too fast. Always wear a face shield or safety glasses.

There is another problem when running an engine in the shop. A poisonous gas called carbon monoxide comes out of an engine's exhaust. Carbon monoxide cannot be seen; it has not color or smell. Make sure large doors or exhaust vents are open before you start an engine, Figure 1-2. Always check with your teacher before you start up an engine. Follow these rules:

1. Get permission from your instructor before starting an engine.

Figure 1-2. An exhaust vent must be used when running an engine inside.

2. Check fuel line for possible leaks.
3. Vent exhaust to the outside of the building, and make sure there is ventilation when you run an engine.
4. Keep your head and hands away from moving parts.
5. Do not run an engine higher than its rated rpm (revolutions per minute).
6. When running an engine at high speed, wear face and ear protection.

USING COMPRESSED AIR

Many shops have compressed air in an air line. It can be dangerous if not used correctly. Follow these rules when you use compressed air:

1. Check all hose connections before turning on the air.

2. Hold the air hose nozzle to prevent it from slipping while turning air on or off.
3. Do not lay the hose down while there is pressure in it. It might whip around and strike someone.
4. Do not use air to dust off hair or clothing or to sweep the floor.
5. Wear safety glasses when using an air hose.
6. Never aim an air nozzle at another person.

FIRES

Many things in the shop can catch fire. You should know how to prevent and fight a fire. Three things must be present to have a fire: oxygen, fuel and heat. To put out a fire, you must remove one of these. Every shop should have a fire extinguisher, Figure 1-3. Find out where the

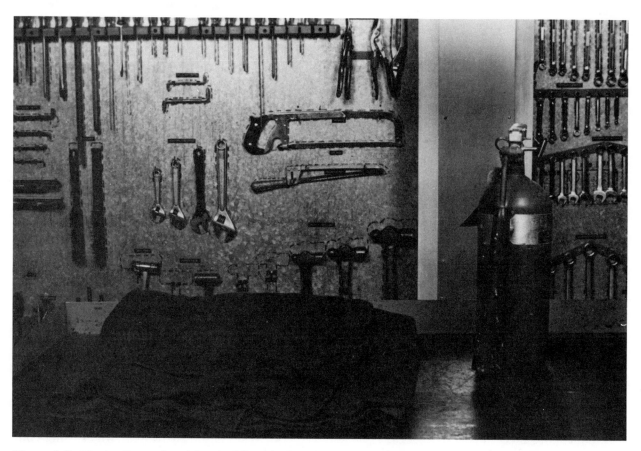

Figure 1-3. Have a fire extinguisher and fire blanket close by when you are working on an engine.

Figure 1-4. Gasoline must be stored in a safety can. (Eagle Manufacturing Co.)

Figure 1-5. Rags containing oil, gasoline or solvent must be kept in a covered metal container. (Eagle Manufacturing Co.)

shop fire extinguisher is and how to use it to put out a fire. To prevent a fire follow these rules:

1. Store flammable liquids in a fireproof room or cabinet.
2. Store gasoline in a safety can, Figure 1-4. A safety can has a top which is held closed by a spring so that gasoline vapors cannot escape from it.
3. Keep rags used to wipe up oil, gasoline, paint, or solvents in metal safety containers, Figure 1-5.
4. Keep the tops of oil cans clean.
5. Drain fuel from an engine before storing the engine in the shop.

NEW TERMS

cleaning equipment: Equipment used to clean small engine parts.
compressed air: Air under pressure.
fire extinguisher: A device designed to put out a fire.
fire prevention: Observation of safety measures to keep fires from starting.
hand tools: Tools that do not use any power except hand power.
power tools: Tools powered by electricity, compressed air or hydraulic fluid.

SELF CHECK

1. List two possible dangers when using hand tools.
2. How should sharp-edged tools be passed to someone?
3. What should you do before turning on compressed air?
4. What should you wear when using compressed air?

5. What should you wear when using cleaning equipment?
6. What can happen if an engine runs at too high a speed for too long?
7. Why is carbon monoxide dangerous?
8. How can you get rid of carbon monoxide?
9. What three things must be present to have a fire?
10. Where is your shop fire extinguisher?

DISCUSSION TOPICS AND ACTIVITIES

1. Examine your home garage for any of the hazards discussed in the chapter. Eliminate any hazards you find.
2. Pass a safety test based on the hazards in your school small-engine shop.

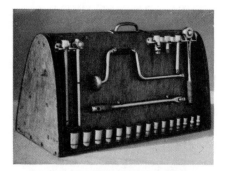

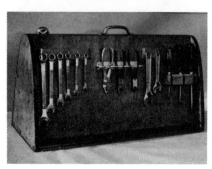

unit 2
using tools

Every small engine repair job requires the use of tools. These tools may be divided into two groups: hand tools and power tools. Hand tools get their name from the fact that the hand supplies the power to operate them. Power tools however, are operated by electricity or air hydraulic power. In this unit we will study the most common small tools you will use in repairing small engines.

LET'S FIND OUT: When you finish reading and studying this unit you should be able to:
1. Identify types of wrenches used to repair small engines.
2. Identify the common types of screwdrivers used on small engines.
3. Describe the common types of hammers and pliers used in the shop.
4. Describe the metalworking tools used for small-engine repair.
5. Identify the common power tools used to repair a small engine.

HAND TOOLS

Wrenches

Many small-engine parts are fastened together with bolts and nuts. Wrenches are tools designed to tighten or loosen bolts and nuts. Since there are many different sizes of bolts and nuts, wrenches must be made in many different sizes.

The size of a wrench is determined by the size of the nut or bolt on which it fits. The wrench shown in Figure 2-1 has 11/16 stamped on it. This

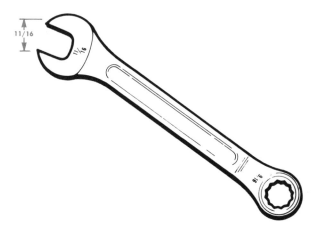

Figure 2-1. The size of a wrench is stamped on the handle or head.

means that the opening of the wrench is 11/16 inch across the flats and that it will fit on a bolt head or nut that is 11/16 inch across the flats.

Wrench sizes are given in the metric system or in the U.S. customary system. Both of these systems are explained in the chapter on measurement. Metric wrenches have sizes such as 10mm, 11mm and 12mm. Wrenches made under the U.S. customary system have fractional sizes such as 5/16″, 3/8″, 7/16″, 1/2″ and 9/16″. In the past, American engines had only customary sized bolts and nuts, while foreign engines used the metric sizes. A mechanic needed two sets of wrenches to work on foreign and American motorcycles. American engine manufacturers, like other American industries, are changing slowly to the metric system. Before long only metric-sized wrenches will be necessary.

Open-End Wrench. An open-end wrench has an opening at the end to allow the wrench to be placed on a bolt or nut, Figure 2-2. Most open-end wrenches have different-sized openings on each end. An open-end wrench should not be used to loosen extremely tight bolts or nuts because it can easily slip off.

Box-End Wrench. A box-end wrench is designed to fit all the way around a bolt or nut, Figure 2-3. Box-end wrenches permit the mechanic to apply more force without the danger of the wrench slipping off the nut or bolt. Like other wrenches, box-end wrenches come in a number of sizes.

Combination Wrench. A combination wrench has a box-end wrench at one end and an open-end at the other, Figure 2-4. A set of combination wrenches provides the advantages of both types. The open-end side is used when space is limited; the box-end is used to finish tightening or to begin loosening. Combination wrenches usually are the same size at both ends.

Socket Wrench. A socket wrench, Figure 2-5, is similar to a box wrench in that it goes all the way around the bolt or nut. Socket wrenches are different in that they can be removed from the han-

Figure 2-4. A combination wrench has a box and an open end. (Snap-On Tools Corporation)

Figure 2-2. An open-end wrench has an opening to allow placement on a bolt or nut. (Snap-On Tools Corporation)

Figure 2-5. A socket wrench goes all the way around a nut or bolt. (Snap-On Tools Corporation)

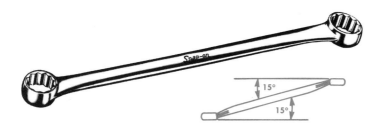

Figure 2-3. A box-end wrench fits all the way around a nut or bolt. (Snap-On Tools Corporation)

dle. Sockets are made in all the U.S. customary and the metric sizes to fit any size bolt or nut. A number of different-sized socket wrenches can be used with one handle.

Sockets are attached to their driving handles and attachments by a square hole at the end of the socket. Socket sets are made with different-sized square drive holes. For small bolts and nuts, such as engine shrouds or sheet metal, socket sets with a 1/4-inch square drive are useful. For general purpose work, a 3/8 drive set is popular. Heavier work requires a 1/2-inch drive socket set.

Besides having different opening sizes and drive sizes, sockets have different numbers of points and different lengths. The socket shown in Figure 2-6 has six points or corners that hold the nut or bolt. The socket shown in Figure 2-7 has eight points. The fewer points, the more strength the socket has, but the harder it is to slip over a bolt or nut.

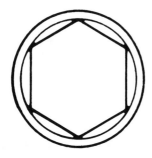

Figure 2-6. A socket with six points.

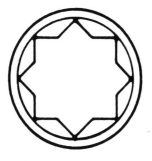

Figure 2-7. A socket with eight points.

Figure 2-8. A deep socket is used for long bolts or studs. (Snap-On Tools Corporation)

The socket shown in Figure 2-8 is an example of a long or deep socket. It is useful for driving nuts on long bolts and studs or working in a deep hole.

A large number of handles and attachments is available to drive socket wrenches. A group of handles and attachments is shown in Figure 2-9. The sockets, handles and attachments must all have the same size drive.

The handle shown in Figure 2-10 is a ratchet. A socket wrench is attached to the square drive on the ratchet handle. The socket is then placed over a bolt or nut that is tightened or loosened by rotating the socket handle. The ratchet has a freewheeling or ratchet mechanism inside it that will allow it to drive the nut in one direction and then allow the handle to move freely in the other direction without driving the nut. This permits fast work in a small space because the socket does not have to be removed from the nut each time it is turned. A lever on the ratchet handle allows the mechanic to choose the direction in which the ratchet will drive.

Torque Wrench. When small-engine parts are reassembled after repair, the bolts and nuts must be tightened to exactly the correct degree. A special type of socket handle called a torque wrench is used for this purpose. A torque wrench measures the force applied to turn a bolt or nut. This force is called torque.

There are many types of torque wrenches. One popular type shown in Figure 2-11 uses a beam and pointer assembly. During tightening, the beam on the wrench bends as the resistance to turning increases. The torque is shown on a scale near the handle. Another type of torque wrench is shown in Figure 2-12. This one has a ratchet-drive head. An adjustable handle and scale allow the

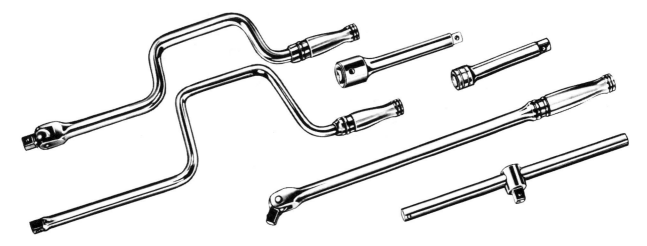

Figure 2-9. Socket wrenches are used with different handles and attachments. (Snap-On Tools Corporation)

Figure 2-10. A ratchet will drive a socket wrench in either direction. (Snap-On Tools Corporation)

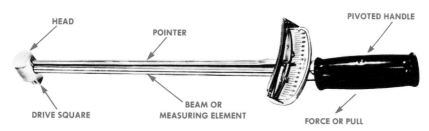

HEAD

POINTER

PIVOTED HANDLE

DRIVE SQUARE

BEAM OR
MEASURING ELEMENT

FORCE OR PULL

Figure 2-11. A torque wrench allows tightening to the correct degree. (Ammco Tools, Inc.)

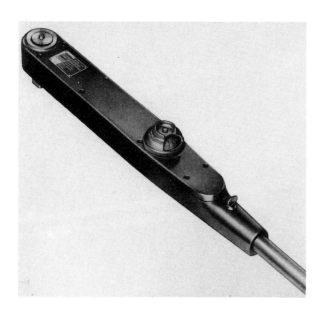

Figure 2-12. Some torque wrenches have a ratchet drive. (Ammco Tools, Inc.)

mechanic to adjust the wrench to a certain torque setting. When that torque is reached, a clicking signal warns the mechanic that the bolt or nut is tightened to that torque.

Adjustable Wrench. The adjustable wrench shown in Figure 2-13 is made to adjust to bolts and nuts of different sizes. These wrenches come in lengths from about four to about 20 inches. The larger the wrench is, the larger the opening can be. For example, a six-inch adjustable wrench will open 3/4 inch wide while a 12-inch wrench will open 1 5/16.

There is a right and a wrong way to use an adjustable end wrench, as shown in Figure 2-14. When tightening, the wrench must be placed on the bolt or nut so that the stress is placed on the stationary jaw (the jaw that does not move). If used incorrectly, the wrench can slip, hurting you and damaging the nut or bolt. In addition, the adjustable jaw of the wrench can be damaged.

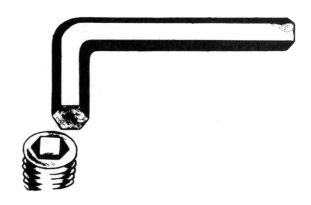

Figure 2-15. A hex-head (allen) screw and wrench.

Hex Wrench. Many drive pulleys on small engines are held in place with hex-head screws. These screws require special wrenches. An example of a hex-head screw and wrench is shown in Figure 2-15. Hex-head screws and wrenches are often called allen screws and allen wrenches. Hex wrenches are available in sets of different sizes, with the size determined by the size of the screw in which they fit. They are made in U.S. customary sizes such as 3/32 and 1/8, or in metric sizes such as 4mm, 5mm and 6mm.

Screwdrivers

Many engine components are held together with screws. A screwdriver is a tool used to turn or drive a screw. It should never be used as a pry bar.

Figure 2-13. An adjustable wrench adjusts to different sizes. (Snap-On Tools Corporation)

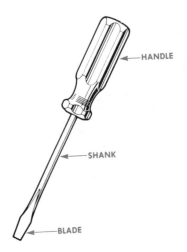

Figure 2-16. Parts of a screwdriver.

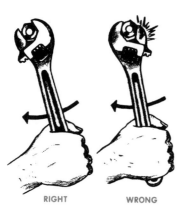

Figure 2-14. Always put the stress on the stationary jaw. (General Motors Corporation)

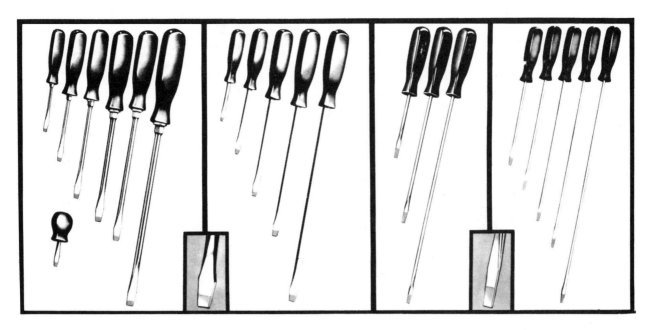

Figure 2-17. A flat-blade screwdriver is used with slotted screws. (Snap-On Tools Corporation)

The parts of a screwdriver are shown in Figure 2-16. The screw is driven by the blade on the screwdriver. The shank connects the blade to the handle. Some screwdrivers have handles made from wood, but plastic is more common. Shanks are made in different lengths and shapes. Some shanks are round; others are square. As we will see below, screwdrivers have different types of blades to drive different types of screws.

Flat-Blade Screwdriver. Screwdrivers used to drive a screw with a straight slot in the top are called flat-blade screwdrivers, Figure 2-17. The length of these screwdrivers sometimes is given by the length of the shank and sometimes by the overall length. Care must be taken to use a screwdriver that fits the slot in the screw properly. The blade should fit into the slot snugly. Otherwise the head of the screw may be damaged.

Phillips Screwdriver. A screw with crosscut slots is driven by a phillips screwdriver, Figure 2-18. A phillips screwdriver is not likely to slip out of the screw. These screwdrivers come in sets of various lengths and blade sizes. The blade sizes are based on a numbering system from 0 (smallest) to 6 (largest).

Figure 2-18. A phillips screwdriver is used with phillips screws.

Pliers

Pliers are tools made for gripping things that wrenches or screwdrivers do not fit. Pliers are never used to drive a nut, bolt or screw. The pliers shown in Figure 2-19 are called slip-joint pliers because there is a slip joint where the two jaws are attached. The slip joint allows the pliers to have two jaw openings: one for holding small objects and one for holding larger objects. These pliers are used for pulling out pins, bending wire, and removing cotter keys. Slip-joint pliers come in many different sizes. They are grouped by their overall length.

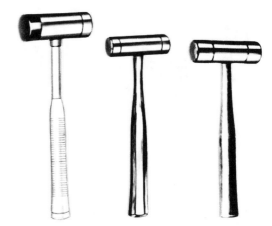

Figure 2-21. Soft-face hammers protect engine parts. (Snap-On Tools Corporation)

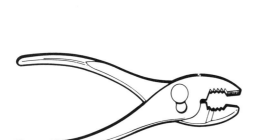

Figure 2-19. Slip-joint pliers are used to hold small objects.

Hammers

Many repair jobs require the use of a hammer. The hammer shown in Figure 2-20 is a ball peen. It has a hardened steel head and is used to drive punches and chisels. Since the ball peen has a hardened head, it should never be used to hammer on a small-engine part. The part could easily be dented or otherwise damaged. Ball peen hammers are made in different sizes and are

grouped by the weight of the head. Small ones weigh as little as four ounces and big ones weigh more than two pounds.

When a small-engine part must be hammered, care must be taken not to damage it. A number of hammers, Figure 2-21, are made with heads softer than small-engine parts. The common soft-face hammer heads are brass, rubber, plastic and rubber-covered steel. The mechanic must choose the correct soft-face hammer for a particular job.

Metalworking Tools

Many small-engine repair jobs require the use of metalworking tools to cut or shape metal. Files, hacksaws, chisels, twist drills, taps, and dies are metalworking tools.

File. A file, Figure 2-22, is a hardened-steel tool used to polish, smooth or shape by removing bits of metal from the object. The face of the file has

Figure 2-20. A ball peen hammer is used to drive punches and chisels. (Snap-On Tools Corporation)

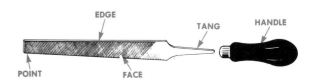

Figure 2-22. A file is used for forming and smoothing surfaces.

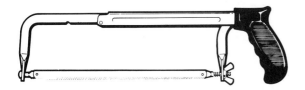

Figure 2-23. A hacksaw is used to cut metal.

Figure 2-24. A chisel is used to cut or shape metal.

Figure 2-25. Twist drills are used to cut holes.

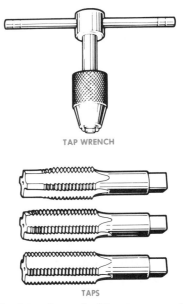

TAP WRENCH

TAPS

Figure 2-26. A tap is used with a tap wrench to make or repair inside threads.

rows of cutting teeth. Files are made in different lengths, measured from the tip to the heel. A pointed end, the tang, is shaped to fit into the handle. A handle must always be attached when filing to protect the mechanic from the sharp tang.

Hacksaw. A hacksaw, Figure 2-23, is a saw designed to cut metal. Mechanics use a hacksaw to cut exhaust pipes and other metal parts that are made during a repair job. The hacksaw may have a rigid frame or an adjustable frame which can take hacksaw blades of different lengths. Hacksaw blades are made in different lengths and with different numbers of teeth.

Chisel. A chisel is a bar of hardened steel with a cutting edge ground on one end. It is used with a hammer to cut or shape metal. The chisel shown in Figure 2-24 is the most common. It is called a cold chisel.

Twist Drills. Twist drills, Figure 2-25, mounted or chucked in an electric drill, are used to drill holes. There are three parts to a twist drill. The end of the drill is called a point, the spiral portion the body, and the part that fits in the electric drill motor is the shank. The tapered-shank drill fits in

a special tapered chuck in a drill press or lathe. The straight-shank drill commonly is used in portable drill motors.

Tap. A tap, Figure 2-26, is a metalworking tool designed to make or repair inside threads. A tap of the correct size is installed in a holding tool called a tap wrench. The tap is then driven into a hole either to make new threads or to repair damaged ones. Taps come in a variety of sizes.

Die. A die, Figure 2-27, is a metalworking tool used to repair or make outside threads. A die is

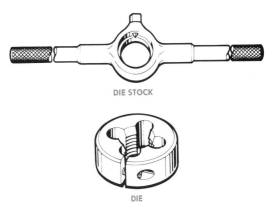

DIE STOCK

DIE

Figure 2-27. A die is used with a die stock to make or repair outside threads.

used in a tool called a die stock. The die is then driven down over the part to make new threads or repair damaged ones. Like taps, dies come in a variety of sizes.

POWER TOOLS

Power tools are tools operated by electricity and air hydraulic power. There are several power tools used in small-engine repair. The most common are the electric wrench, electric drill and air impact wrench.

Electric Wrench

An electric wrench, Figure 2-28, has an electric motor operated by a trigger on the handle. Special heavy-duty sockets are attached to a socket drive at the front of the wrench. Holding down the trigger spins the drive and socket. A reversing switch allows the mechanic to loosen as well as tighten bolts and nuts. The main advantage of the electric wrench is speed. Its motor drives a socket much faster than it can be driven by hand. Electric wrenches are especially useful for disassembling parts held together with many bolts and nuts, such as engines and transmissions.

Electric Drill

An electric motor operated by a trigger on the handle drives a chuck at the front of the electric drill, Figure 2-29. This power tool is used not only to drill holes, but also to drive engine cylinder hones and deglazers.

Figure 2-29. Twist drills are driven by an electric drill. (Black & Decker Co.)

Figure 2-28. An electric wrench drives bolts or nuts electrically.

Figure 2-30. A bolt or nut which is difficult to loosen may be removed with an impact wrench. (K-D Tools)

Air-Impact Wrench

An air-operated wrench, Figure 2-30, is connected to an air line. Pulling the trigger causes the air to rotate a socket attached to the drive on the wrench. A reversing switch allows the mechanic to loosen as well as tighten. Many air wrenches are designed with an impact feature. An impact wrench not only drives the socket but also vibrates or impacts it in and out. The force of the impact helps to loosen a bolt or nut that is difficult to remove.

NEW TERMS

adjustable wrench: A wrench designed to adjust to different sizes of bolts and nuts.

air-impact wrench: A wrench powered by compressed air.

chisel: A bar of hardened steel with a cutting edge ground on one end. It is driven with a hammer to cut metal.

combination wrench: A wrench with a box end at one end and an open end at the other.

die: A tool used to cut external threads.

electric drill: A drill powered by electricity.

electric wrench: A wrench powered by electricity.

file: A hardened steel tool with rows of cutting edges used to polish, smooth or shape by removing metal.

flat blade screwdriver: A screwdriver with a blade or tip made to drive common slotted screws.

hacksaw: A saw for cutting metal.

hex-head wrench: A wrench used to tighten or loosen allen, hex-head or hollow-head screws.

open-end wrench: A wrench with an opening at the end to allow it to be positioned on the bolt or nut.

phillips screwdriver: A screwdriver with a point on the blade or tip, used for driving phillips-head screws.

pliers: A tool designed for gripping objects that wrenches or screwdrivers will not fit.

power tool: Any tool powered by electricity and compressed air.

screw: A threaded fastener that fits into a threaded hole in an automotive component.

socket handles and attachments: Tools used to drive socket wrenches.

socket wrench: A wrench that fits all the way around a bolt or nut and is made to be detached from a handle.

tap: A tool used to cut internal threads.

torque wrench: A wrench designed to tighten bolts or nuts to the correct tightness or torque.

twist drill: A hardened cutting tool made to cut or drill a hole.

SELF CHECK

1. What are wrenches used for?
2. In what two ways are wrench sizes specified?
3. What does *10mm* mean when stamped on a wrench?
4. Describe an open-end wrench.
5. Describe a box-end wrench.
6. Describe a combination wrench.
7. Explain how a socket wrench differs from other types of wrenches.
8. Describe what a torque wrench is and why it is used.
9. Describe a flat blade screwdriver.
10. When are soft-face hammers used?
11. For what are pliers used on small engines?
12. List four metalworking tools.
13. List three power tools used in the shop.

DISCUSSION TOPICS AND ACTIVITIES

1. Study the tools in your school small-engine shop. How many can you name?
2. Make a list of the hand tools in your home garage.

unit 3

using fasteners

A small engine is assembled and held together with a number of small devices such as bolts, nuts and screws. These parts are called fasteners because they are used to fasten or hold parts together. Every repair job involves the use of fasteners. In this unit we will study the most common fasteners you will be using.

LET'S FIND OUT: **When you finish reading and studying this unit, you should be able to:**
1. **Define the term** *fastener*.
2. **Explain the purpose of fasteners.**
3. **Identify the common threaded fasteners.**
4. **Explain how threads are described or designated.**
5. **Identify the common non-threaded fasteners.**

THREADED FASTENERS

Threaded fasteners use the wedging action of a spiral groove or thread to clamp two parts together. The common types of threaded fasteners are screws, bolts, studs and nuts.

Screw

A screw fits into a threaded hole in an automotive component. The screw is turned or driven into the threaded hole to hold or clamp two parts together. The cap or hex-head screw, shown in Figure 3-1, is the most common screw used in small-engine work. The cap screw is driven or turned with a common wrench such as the box, open-end, combination or socket wrench.

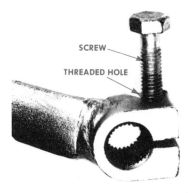

SCREW

THREADED HOLE

Figure 3-1. A hex-head or cap screw fits into a threaded hole.

18

Bolt

A bolt is used with a nut instead of a threaded hole, Figure 3-2. The only difference between a bolt and a screw is in their use: a screw is used in a threaded hole, and a bolt is used with a nut. Many bolts look like the cap screw shown earlier. When bolts are tightened or loosened, two wrenches normally are required, one to drive the bolt and one to keep the nut from turning.

Figure 3-2. A bolt is used with a nut to clamp parts together.

Stud

A stud has threads at both ends, Figure 3-3. One end of the stud fits into a threaded hole in a part. Another part fits over the stud, and the two parts are clamped together using a nut. Studs often are used where the positioning of a part is important. A stud may be threaded all along its length, but it is more common to find threads only on the ends.

Figure 3-3. A stud has threads at both ends.

Nut

Nuts have an internal thread and are used with bolts and studs. Most small-engine nuts are hex-shaped like that shown in Figure 3-4. They may be driven with box-end, open-end or socket wrenches.

Figure 3-4. A hex nut is used with bolts and studs.

WASHERS

Washers are used with bolts, screws, studs and nuts. A flat washer, shown in Figure 3-5, often is used between a nut and an automotive component, or under the head of a screw or bolt, to spread the clamping force over a wider area. It also prevents a machined surface from being scratched as the bolt head or nut is tightened.

Figure 3-5. A flat washer is used with a nut, screw or bolt.

Figure 3-6. A lock washer prevents parts from working loose.

Figure 3-7. Fine- and coarse-threaded fasteners must not be used together.

Washers also may be used to prevent fasteners from vibrating or working loose. These washers are called lock washers, Figure 3-6. A lock washer has a sharp edge that will dig into a fastener or component surface. This prevents the fastener from working loose.

THREAD SIZES AND DESIGNATIONS

Several types of threads are used in threaded fasteners. Since one kind of thread cannot be used with another, the mechanic must understand the different thread systems. In this section we will present the types of threads used in small-engine work.

Unified System Thread Designation

Small engines manufactured in the United States have, until very recently, used only the Unified System threads. This system includes two types of threads, coarse and fine. Unified National Coarse threads are designated by UNC, and Unified National Fine threads are designated by UNF. Figure 3-7 shows the difference between the two. Coarse has fewer threads per inch than fine. The difference is easily seen, and they must not be mixed. Trying to tighten a coarse-thread cap screw into a hole with fine threads, for example, will damage the threads. Coarse threads are used in aluminum parts because they provide greater holding strength in soft materials. Fine threads are used in many harder materials such as cast iron and steel.

Unified threaded fasteners are designated by size also. The sizes of most bolts, screws, studs and nuts are based upon fractions of an inch. A typical designation is 1/2-20UNF. The 1/2 represents a bolt, screw or stud thread diameter of 1/2 inch. When used with a nut, the fraction refers to the bolt size it fits. The 20 refers to the number of threads per inch, and the UNF stands for Unified National Fine threads. All 1/2-inch National Fine fasteners have 20 threads per inch. A coarse-thread fastener 1/2 inch in diameter is designated 1/2-13UNC and has 13 threads per inch.

Metric System Thread Designation

Small engines manufactured using the metric system have fasteners with metric threads. Metric and Unified National fasteners cannot be mixed. An example of a designation for metric threads is M12 x 1.75. The M indicates that the fastener has metric threads. The number following the M refers to the outside diameter of a bolt, screw or stud, or to the inside diameter of a nut. This measurement is in millimeters. The last number, separated by the sign x, gives the pitch, that is, the distance in millimeters between each of the threads.

There are fine and coarse metric threads, and the pitch number is used to distinguish them. A fine-thread metric bolt, for example, may be indicated by M8 x 1.0. A bolt of the same diameter with a coarse thread is designated M8 x 1.25. The larger pitch number indicates a wider space between threads.

NON-THREADED FASTENERS

Besides threaded fasteners, there are other devices for holding small engine parts together. The most common non-threaded fasteners are dowel pins and keys.

Dowel Pin

A dowel pin is a round pin that fits into a drilled hole to position two parts that fit together. It is usually quite small and requires special pliers or a punch and hammer for removal. Dowel pins may be straight or tapered, solid or split.

Key

A key is a small hardened piece of metal used with a gear or pulley to lock it to a shaft. One half of the key fits into a keyseat on the shaft, while the other half fits into a slot called a keyway on the pulley or gear. The keyseat, keyway and key are shown in Figure 3-8. Gears and pulleys are held on an engine crankshaft with keys.

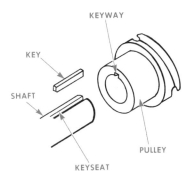

Figure 3-8. A key is used to lock a pulley to a shaft.

NEW TERMS

bolt: A threaded fastener used with a nut to hold small engine parts together.

dowel pin: A round metal pin that fits into drilled holes to position two mating parts.

hammer: A tool used to drive or pound on an object. Hammers for automotive use may have hard or soft heads.

keys: A small hardened piece of metal used with a gear or pulley to lock it to a shaft.

non-threaded fastener: A device used to hold small engine components together without the use of threads.

nut: A small fastener having internal threads, used with bolts and screws.

stud: A fastener with threads at both ends.

thread designation: A system used to indicate the size of the threads on threaded fasteners.

threaded fasteners: A device that uses threads to hold automotive components together.

washer: Used with bolts, screws, studs and nuts to distribute the clamping force and to prevent fasteners from vibrating loose.

SELF CHECK

1. Define the term *fastener*.
2. What is the purpose of fasteners?
3. How does a threaded fastener work?
4. Describe the difference between a bolt and a screw.
5. Where are studs used?
6. What is the purpose of a washer?
7. What does the *UNC* in a thread designation mean?
8. What does *UNF* mean in a thread designation?
9. What does the *M* mean in a metric thread designation?
10. Describe two types of non-threaded fasteners.

DISCUSSION TOPICS AND ACTIVITIES

1. Make a list of all the fasteners you can find in the shop.
2. Design and build a storage tray for the fasteners you have in your home garage.

unit 4

using measuring systems and tools

The small engine and each of its parts is made with a great deal of precision. The need for precision does not stop at the end of the assembly line. Precision is an important part of service. The mechanic is furnished with specifications, tolerances and clearances which must be correct or the engine will not work properly. To use these specifications, a mechanic must be able to measure precisely. The purpose of this unit is to explain the two measuring systems the mechanic must use and to present the precision measuring tools used in small-engine service. In later units we will see how these tools are used to measure engine parts.

LET'S FIND OUT: When you finish reading and studying this unit, you should be able to:
1. Explain the customary measuring system and use the customary units.
2. Explain the metric system of measurement and use the metric units.
3. Convert between the customary and metric system units.
4. Identify and describe the purpose of the common measuring tools.
5. Read the measuring scales on precision measuring tools

THE CUSTOMARY SYSTEM OF MEASUREMENT

The customary system of measurement is one of two main measuring systems used in the world today. Originally it was the most important system. It is still the most common system used in the United States.

The customary system may be divided into units of length-measurement and units of weight or mass measurement. These units are shown in Figure 4-1. For our purposes, units of weight or mass are not very important to engine service. Units of length measurement are very important because most specifications the small-engine mechanic uses are measurements of length.

ENGLISH SYSTEM MEASUREMENT UNITS

UNITS OF LENGTH MEASUREMENT		
12	INCHES	= 1 FOOT
3	FEET	= 1 YARD
51/2	YARDS	= 1 ROD

UNITS OF WEIGHT MEASUREMENT		
16	DRAMS	= 1 OUNCE
16	OUNCES	= 1 POUND

Figure 4-1. Customary system units.

One foot may be divided into 12 inches. Small-engine specifications often are written in terms of a part of an inch. Two systems are used in subdividing an inch: fractions and decimals.

Fractions

1 in.   (one inch)
1/2 in.   (one half inch)
1/4 in.   (one quarter inch)
1/8 in.   (one eighth inch)
1/16 in.   (one sixteenth inch)
1/32 in.   (one thirty-second inch)
1/64 in.   (one sixty-fourth inch)

A typical ruler is divided into these units. The smallest division of a rule is 1/64 of an inch, about equal to the thickness of a fingernail. Often it is necessary to make measurements much smaller than 1/64 of an inch. This is made possible by the decimal system of dividing an inch. With this system, the inch is divided by ten, this in turn by ten, and so on as shown below:

Decimals

1.0 in.   (one inch)
0.1 in.   (one-tenth inch)
0.01 in.   (one-hundredth inch)
0.001 in.   (one-thousandth inch)
0.0001 in.   (one-ten-thousandth inch)

Many small-engine components must be measured to within one thousandth of an inch (0.001). Some specifications require a mechanic to measure as closely as one-ten-thousandth of an inch (0.0001). By comparison, a human hair is about three thousandths of an inch (0.003) thick.

THE METRIC SYSTEM OF MEASUREMENT

Up to 1790, each country and area used a different type of measuring system, making trade very difficult. The French government realized that a single, simple measuring system would benefit all countries. Today this metric system is used by most of the countries in the world. The United States is one of the last countries to adopt it. Since the changeover from the customary to the metric system will take a long time, a small-engine mechanic must be familiar with both systems.

The French also realized that it would not be practical to measure very small and very large things with the same units. For example, the measurement of a human hair and a soccer field require different units. The problem was solved very well. In the metric system, larger and smaller units are defined from the basic meter in decimal steps. Their names are formed by adding one of the prefixes shown in Figure 4-2 to the word "meter" (in French, "metre").

Using these prefixes, it is easy to describe lengths of any size. For example, a human hair may be measured in millimeters, the length of a soccer field in meters and the distance from the earth to the sun in gigameters.

Another convenient part of the metric system is that each prefix can be abbreviated with a symbol, as shown in Figure 4-2. The symbol for the meter is m. To this symbol can be added the prefix symbol. For example, 1,000 meters may be written 1,000 m, or, since a kilometer is equal to 1,000 meters, it may be written 1 kilometer or simply 1 km. Similarly, one thousandth of a meter may be written 1 millimeter or simply 1 mm.

NUMBER OF METERS	PREFIX	SYMBOL
1 000 000 000	gigameter	Gm
1 000 000	megameter	Mm
1 000	kilometer	km
100	hectometer	hm
10	decameter	dam
1	meter	m
0.1	decimeter	dm
0.01	centimeter	cm
0.001	millimeter	mm

Figure 4-2. Units, prefixes and symbols used in the metric system.

CONVERTING BETWEEN CUSTOMARY AND METRIC SYSTEM UNITS

Since both the customary and the metric systems are used in this country, frequently it is necessary to convert units from one system to another. A mechanic may find a specification written in millimeters but have measuring tools that measure in thousandths of an inch. Conversion charts provide a means of switching from one system to another. A conversion chart lists units of one system in one column and their equivalents from the other system beside them in a second column. A conversion chart is located in the appendix of this book.

It also is possible to make the changes yourself, by multiplying by a number called a conversion factor. The conversion factors are shown in Figure 4-3. If a specification is written as 10 mm (ten millimeters) and you want to know what this is in inches, you use the conversion factor as follows:

Multiply 10 mm by the conversion factor in Figure 4-3, .03937:

10 mm × .03937 = .3937 inches

10 mm is equal to .3937 inch

In order to change 5 miles into kilometers, multiply 5 miles by the conversion factor in Figure 4-3, 1.609:

5 miles × 1.609 = 8.045 kilometers

5 miles is equal to 8.045 kilometers

USING MEASURING TOOLS

In this section we will examine the measuring tools or instruments commonly used by the small-engine mechanic. In each case, both the customary and the metric measuring tool will be presented.

Rule

The rule or ruler is the simplest of all measuring tools. A rule is a flat length of wood, paper, plastic or metal divided, or graduated, into a number of spaces. Rules using the customary system of measure are usually six or twelve inches long, Figure 4-4. Each inch is divided into several parts. Some rules divide the inch into 8, 16 or 32 parts. Precision machinist rules go to 1/64 inch. This is the smallest division of an inch that can be read with the unaided eye.

The rule shown in Figure 4-5 is divided into metric units. Metric rules commonly are subdivided into centimeters and millimeters. Some metric rules are further divided into .5 millimeter spaces. Reading the metric rule is easier than reading the customary rule because there is no need to add fractions.

Another metric rule is shown in Figure 4-6. This rule is 100 centimeters long. Every small mark is one millimeter. Every 10th mark is larger and is equal to one centimeter. Alongside the

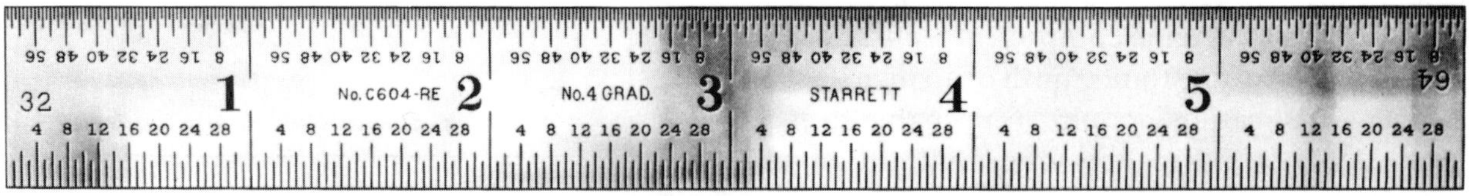

TO FIND		MULTIPLY	X	CONVERSION FACTORS
MILLIMETERS	=	INCHES	×	25.40
CENTIMETERS	=	INCHES	×	2.540
CENTIMETERS	=	FEET	×	32.81
METERS	=	FEET	×	.3281
KILOMETER	=	FEET	×	.0003281
KILOMETER	=	MILES	×	1.609
INCHES	=	MILLIMETERS	×	.03937
INCHES	=	CENTIMETERS	×	.3937
FEET	=	CENTIMETERS	×	30.48
FEET	=	METERS	×	.3048
FEET	=	KILOMETERS	×	3048.
YARDS	=	METERS	×	1.094
MILES	=	KILOMETERS	×	.6214

Figure 4-3. Converting from customary to metric units.

Figure 4-4. Customary system rule. (L. S. Starrett Co.)

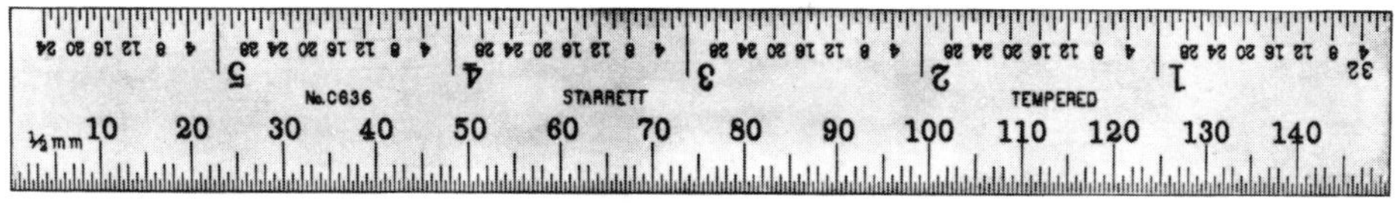

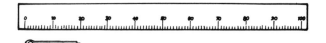

Figure 4-5. Metric system rule. (L. S. Starrett Co.)

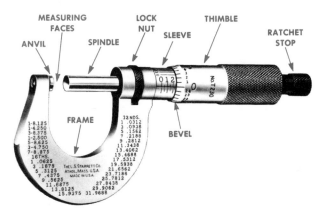

Figure 4-6. Measuring with a metric rule.

metric rule in Figure 4-6 is a cotter pin. The length of the cotter pin may be measured as 20 millimeters or two centimeters. Remember that the relationship between a centimeter and a millimeter is the same as between a dime and a penny. (Ten millimeters equals one centimeter).

Outside Micrometer

Outside micrometers, commonly called "mikes," provide the most precise measurements required for general engine service. Micrometers are made in different sizes and shapes and for a number of special purposes. For most measuring jobs, however, the standard outside micrometer shown in Figure 4-7 is used. These micrometers are marked with either customary or metric measuring units.

Figure 4-7. Parts of a micrometer. (L. S. Starrett Co.)

The basic parts of the micrometer, shown in Figure 4-7, are a frame, anvil, spindle, sleeve and thimble. The measuring surfaces are at the ends of the stationary anvil and the movable spindle. The spindle is actually an extension of a precision-ground screw which threads into the sleeve. The other end of the screw is attached to the thimble. So, turning the thimble moves the spindle toward or away from the anvil.

The item to be measured is placed between the anvil and spindle faces. The spindle is rotated by means of the thimble until the anvil and spindle both contact the item to be measured. The dimension is then found from the micrometer reading shown by the gradations on the sleeve and thimble.

Reading a Customary System Micrometer. The spindle is attached to a screw that is ground to extremely accurate specifications. One revolution of the spindle moves it .025 inch toward or away from the anvil. So, 40 turns of the thimble will move the spindle exactly one inch (40 × .025″ = 1.000″).

A scale on the sleeve is divided into 40 gradations, each equal to .025 inch. So, starting with the spindle against the anvil, and turning the screw out, every revolution of the thimble will uncover one of the divisions on the sleeve. Every fourth division is numbered, starting with the zero mark, when the spindle is against the anvil. The next numbered division is at .100 inch from the closed position. (This is the same as 1/10 of an inch.) The three unnumbered divisions between zero and one are .025, .050 and .075 inch.

The bevel on the front of the thimble also is divided into equal parts. And, since the thimble and spindle travel .025 inch per revolution, there are 25 divisions on the bevel. These divisions make it possible to read the amount of spindle travel for partial revolutions. For instance, a partial revolution from one thimble mark to the next is 1/25 of a revolution and moves the spindle .001

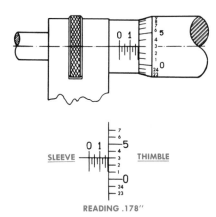

READING .178″

Figure 4-8. A micrometer reading of .178″. (L. S. Starrett Co.)

inch. Reading a measurement taken with micrometers is a simple matter of addition. In other words, add together the last visible numbered division on the sleeve, the unnumbered sleeve divisions and the divisions on the bevel of the thimble.

To help explain, look at the illustration shown in Figure 4-8. The 1 line on the sleeve is visible, representing .100″. There are three additional lines visible, each representing .025″ (3 × .025″ = .075″). Line three on the thimble coincides with the longitudinal line on the sleeve, each line representing .001″ (3 × .001 = .003″). The micrometer reading is .178″ (.100 + .075 + .003). An easy way to remember is to think of the various units as if you were making change from a $10 bill. Count the figures on the sleeve as dollars, the vertical lines on the sleeve as quarters, and the divisions on the thimble as cents. Add up your change and put a decimal point instead of a dollar sign in front of the figures.

The micrometer we have been studying up to this point has a range from 0 to 1 inch. Micrometers are made in many sizes — 1 to 2 inches, 2 to 3 inches and on up to measure large components. Adapters are available for the larger micrometers to give them multiple ranges. These are used exactly as the 0 to 1 inch micrometer.

Reading a Metric System Micrometer. A metric micrometer has the same parts and works in exactly the same way as an customary system

micrometer. The pitch of the spindle screw in metric micrometers is 0.500 millimeters. One complete revolution of the thimble advances the spindle toward or away from the anvil exactly 0.500 millimeters. Two complete revolutions of the thimble move the spindle exactly one millimeter.

The longitudinal line on the sleeve is graduated in millimeters from 0 to 25 and each millimeter is subdivided into 0.5 mm. Therefore, it requires two revolutions of the thimble to advance the spindle one millimeter.

The beveled edge of the thimble is graduated into 50 divisions, from 0 to 50, with every fifth line being numbered. Since a complete revolution of the thimble advances the spindle 0.5 mm, each gradation on the thimble is equal to 1/50 of 0.5 mm or 0.01 mm. Two gradations equal 0.02 mm, etc.

To read a metric micrometer, add the total reading in millimeters visible on the sleeve to the reading in hundredths of a millimeter on the thimble.

For example, refer to Figure 4-9:

The 5 mm mark is visible on the sleeve, representing 5.00 mm.

There is one additional 0.5 mm line visible, representing 0.50 mm.

Line 28 on the thimble coincides with the longitudinal line on the sleeve, each line representing 0.1 mm (28 × .01 mm = 0.28 mm).

The micrometer reading is 5.78 mm.

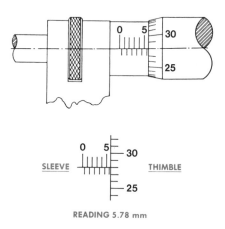

READING 5.78 mm

Figure 4-9. A micrometer reading of 5.78 mm. (L. S. Starrett Co.)

Use and Care of the Micrometer. There really is no trick to reading a micrometer, but there are some things to keep in mind when you are taking the measurement. For instance, there is a right way to hold the micrometer when measuring an object. The most convenient way is with one hand. Insert one finger through the frame and use the thumb and forefinger to turn the spindle, as shown in Figure 4-10. With a little practice, you will find that this will give you the best control over the position of the anvil and spindle.

It is impossible to get a correct measurement with a micrometer unless the anvil and spindle are at right angles to the piece being measured. If they are cocked to one side, you will get an oversize reading. And, if you are measuring a diameter, make sure the spindle and anvil are centered exactly across the diameter; otherwise, the reading will be undersize. To avoid cocking and to assure being on the true diameter, hold the micrometer loosely, and gently turn the spindle down against the workpiece. Rocking the micrometer ever so slightly as you turn the spindle down the last few thousandths will enable you to tell by "feel" alone when the micrometer is square with the part and centered on the diameter.

Probably the most important thing to observe when you are measuring with a micrometer is the amount of force you use to tighten the spindle down onto the part. The spindle and anvil should just contact the part lightly, so there is a slight

drag when the micrometer is moved back and forth. Just keep in mind that a micrometer is a precision tool, not a C-clamp.

If you crank the spindle down too hard, not only will you get an incorrect reading, but you might also distort the frame. Once the frame is distorted, the micrometer is useless. Some micrometers have a ratchet on the end of the thimble. Unless you have a good "feel" for the correct tightness, turn the ratchet instead of the thimble to avoid overtightening.

Sometimes you want to run the spindle in or out in a hurry to measure another workpiece. If you do, hold the frame in one hand and roll the thimble along the other arm or along the palm of the hand.

When the micrometer is not in use, it should be stored in a box in a safe place where tools will not be dropped on it accidentally. Make sure the spindle is backed off slightly, away from the anvil.

As with any other precision tool, a micrometer occasionally should be checked for accuracy. Use a master gage to check the maximum and minimum limits of measurement. For instance, to check a one-inch micrometer, use the master, which is exactly one inch in diameter. Then, run the spindle down gently against the anvil and check for a zero reading. Always make sure the spindle and anvil are clean before checking.

Inside Micrometer

Another very useful measuring tool, an inside micrometer, is shown in Figure 4-11. This tool is especially valuable when boring and honing cylinders. An inside micrometer is used to measure holes. The same micrometer is used to measure many different diameters. As shown in Figure 4-12, measuring rods and spacing collars of different lengths are supplied with the microme-

Figure 4-10. This is how to hold a micrometer.

Figure 4-11. An inside micrometer is used to measure the inside of cylinders. (L. S. Starrett Co.)

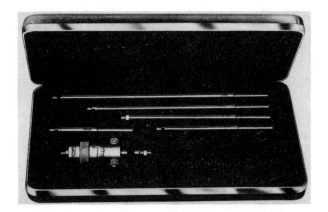

Figure 4-12. An inside micrometer is used with a selection of rods of different lengths. (L. S. Starrett Co.)

Figure 4-13. Set of small hole gages. (L. S. Starrett Co.)

ter. Different ranges of measurement are possible by assembling different rods in the micrometer head.

The scale on the inside micrometer works and is read exactly like that for the outside micrometer. Inside micrometers are made in both customary and metric system scales. While the scales are the same, it will take a little more practice to get an accurate measurement with an inside micrometer. It is easy to get it cocked in the bore and get an incorrect reading. To get an accurate measurement, make sure the micrometer is at right angles to the centerline of the bore. Then, move one end back and forth slightly to get the maximum reading on the scales. It is always a good idea to take two or three additional readings just to check yourself.

Small-Hole Gage

When it is necessary to measure the inside of a hole that is too small for an inside micrometer, a small-hole gage, Figure 4-13, is used. This tool consists of a split sphere, the diameter of which can be changed by means of an internal wedge. The size is changed by turning the handle. The gage is placed into the hole to be measured and adjusted to fit the internal dimension. It is then removed from the hole and an outside micrometer used to measure the diameter of the expanded ball. Small-hole gages are available in sets that cover a range from 3 mm to 12 mm (.125 to .500 inch).

Telescoping Gage

The telescoping gage, Figure 4-14, gets its name from the fact that it consists of a spring-loaded piston which telescopes within a cylinder.

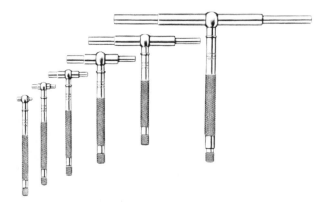

Figure 4-14. Set of telescoping gages. (L. S. Starrett Co.)

It is used with an outside micrometer to measure the inside dimension of a hole. Telescope gages are made in sets, such as that shown in Figure 4-14, to cover a range from very small to very large holes.

The gage is placed into the hole, permitting the spring-loaded piston to expand to the size of the hole. After the proper "feel" is obtained, the handle is turned to lock the piston in position. The exact size of the hole is then found by removing the gage and measuring across the two contacts with an outside micrometer.

Feeler Gage

A feeler gage is a tool used to measure accurately the space between two surfaces, such as breaker points or the gap between the two electrodes on spark plugs. A feeler gage is a flat blade or round wire made to a very precise thickness. The thickness is written on the gage in thousandths of an inch or hundredths of a millimeter. A feeler gage is used by placing it in the space to be measured. If the gage and the space are the same size, the gage will feel tight as it is moved in and out. Feeler gages usually come in sets. A set of metric feeler gages is shown in Figure 4-15. A set of customary system feeler gages is shown in Figure 4-16.

Figure 4-16. Customary feeler gage set. (L. S. Starrett Co.)

Dial Indicator

A dial indicator is a gage that is used to measure the movement, or "play," and the contour, or "runout," of an engine part. The measurement is

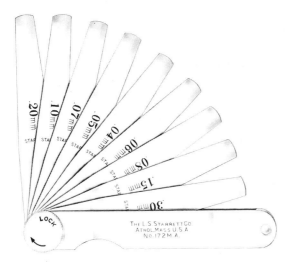

Figure 4-15. Metric feeler gage set. (L. S. Starrett Co.)

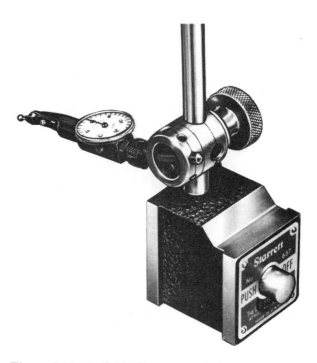

Figure 4-17. A dial indicator attached to a magnetic base. (L. S. Starrett Co.)

shown by a pointer on the face of the gage. The most common type of dial indicator uses a plunger or lever connected to a pointer by a gear built into the instrument. Movement of the plunger is shown by the pointer. The dial indicator is used with a number of attachments that allow it to be mounted on small-engine components. The dial indicator assembly shown in Figure 4-17 has a magnetic base which allows it to be attached to any iron or steel small-engine component. A clamp is used to mount it to an aluminum engine.

Dial indicators used in most small-engine repair operations measure either in thousandths of an inch (0.001″) or in hundredths of a millimeter (0.01 mm). A typical customary dial indicator face is shown in Figure 4-18. The scale is divided into 100 divisions. Each of the divisions represents .001 of an inch. The pointer in Figure 4-18 is pointing to 75 on the scale. This means that the plunger has moved to measure a distance of .075.

When mounting a dial indicator, keep the support arms as short as possible. If the arms are too long, the setup will not be rigid enough, and an inaccurate reading may result. The spring load on the indicator plunger can move the whole indicator assembly.

Mount the indicator in a position that will place the plunger directly against the part. If the anvil is at an angle, the anvil plunger will be subject to frictional drag, causing an incorrect reading. The friction will cause the whole indicator assembly to move, instead of just the anvil and plunger. Always read the dial indicator straight on. Looking at it from the side can cause considerable error. Remember that a dial indicator is a precision instrument like a watch. It must be handled with great care.

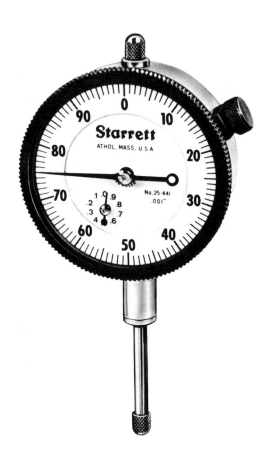

Figure 4-18. A dial indicator reading of .075. (L. S. Starrett Co.)

NEW TERMS

customary measuring system: One of two main measuring systems in use in the world. Most common system used in the United States.

customary system units: Units are based on the yard measurement. The yard is divided into feet and inches.

dial indicator: A gage used to measure movement, or "play," and contour, or "runout," of a small-engine part.

feeler gage: A measuring tool used to measure accurately the space between two surfaces.

inside micrometer: A measuring tool used to measure the size of holes such as small-engine cylinders.

metric measuring system: One of the two main measuring systems in use in the world. Used for a long time in other countries, it is being adopted gradually in the United States.

metric units: Metric system units are based on the meter. Measurements are based on decimal steps of the meter.

outside micrometer: A measuring tool used to measure the outside of an object such as a crankshaft or piston.

rule: The simplest of all measuring tools. It is a flat length of wood, plastic or metal divided into a number of measuring units.

small hole gage: A measuring tool consisting of a split sphere and an internal wedge, used to measure the inside of small holes such as valve guides.

telescoping gage: A measuring tool with a spring-loaded piston that telescopes within a cylinder. It is used to measure the inside of a hole.

SELF CHECK

1. List the divisions of an inch between 1 inch and 1/64 inch.
2. Write the following parts of an inch in numerals: one inch, one-tenth inch, one-hundredth inch, one-thousandth inch.
3. List three prefix symbols used with a meter.
4. Explain how measurements can be converted from one measuring system to another.
5. What is the simplest measuring tool?
6. Describe an outside micrometer and explain its purpose.
7. List the five basic parts of an outside micrometer.
8. Describe the precautions that should be followed when using an outside micrometer.
9. How is the inside of a hole measured?
10. How do telescoping gages work?
11. How is the inside of a small hole measured?
12. Describe a feeler gage and explain how it is used.
13. Describe a dial indicator and explain its purpose.
14. What does the term "play" mean when used with a dial indicator?

DISCUSSION TOPICS AND ACTIVITIES

1. Measure the following objects with customary and metric system rules. Record your results.
   Thickness of a penny
   Diameter of a penny
   Width of your thumb
   Your height
   Length of your shoe
2. Use an outside micrometer to measure the diameter of a hair, the diameter of a paper clip and the thickness of a pencil lead.

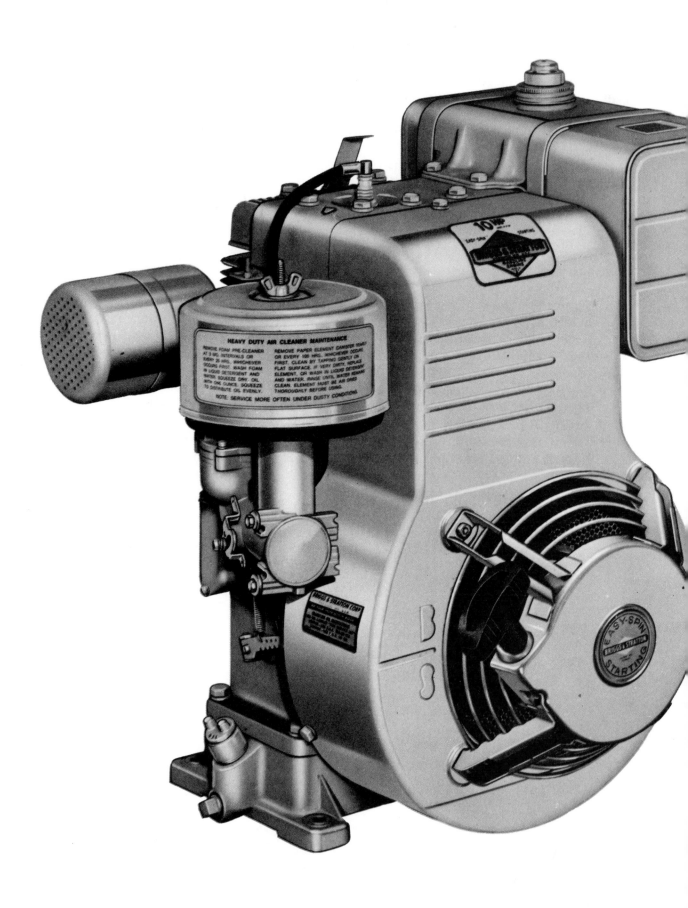

HEAVY DUTY AIR CLEANER MAINTENANCE

REMOVE FOAM PRE-CLEANER
AT 3 MO. INTERVALS OR
EVERY 25 HRS., WHICHEVER
OCCURS FIRST. WASH FOAM
IN LIQUID DETERGENT AND
WATER. SQUEEZE DRY. OIL
WITH ONE OUNCE. SQUEEZE
TO DISTRIBUTE OIL EVENLY.

REMOVE PAPER ELEMENT CANISTER YEARLY
OR EVERY 100 HRS., WHICHEVER OCCURS
FIRST. CLEAN BY TAPPING GENTLY ON
FLAT SURFACE. IF VERY DIRTY, REPLACE
ELEMENT, OR WASH IN LIQUID DETERGENT
AND WATER. RINSE UNTIL WATER REMAINS
CLEAN. ELEMENT MUST BE AIR DRIED
THOROUGHLY BEFORE USING.

NOTE: SERVICE MORE OFTEN UNDER DUSTY CONDITIONS.

part 2

how small engines work

An engine sometimes is called a power plant or a motor. One type of engine uses the explosive power of a mixture of air and gasoline to drive down pistons. The pistons are attached to a crankshaft and force it to turn. The rotating force of the crankshaft is used to do work. We will study small engines that operate on the four-stroke, two-stroke and diesel principles.

Another kind of engine is known by a variety of names: rotary, rotating combustion, or Wankel (after its inventor, Felix Wankel). The Wankel engine does the same thing other engines do. It draws in a mixture of air and fuel, compresses and burns it, but it does it in a different way. A rotor revolves in a chamber. It is connected to a shaft which does the work. We will study this engine too.

33

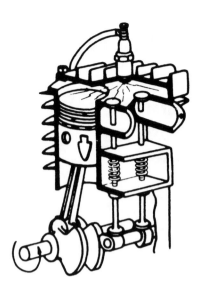

unit 5

fundamentals of the four-stroke-cycle engine

An engine is a machine that changes fuel and air into power. A fuel, such as gasoline, is mixed with air. The mixture of air and gasoline is then burned to create an expanding gas. The expanding gas is used to develop power. In this unit we are going to see how an engine works to develop the power. We will study an engine called the four-stroke cycle. Many small engines are four-stroke-cycle engines.

LET'S FIND OUT: When you finish reading and studying this unit, you should be able to:
1. List the basic parts of an engine.
2. Explain how a basic engine works.
3. List and describe what happens in the four strokes.
4. Name the parts of a small four-stroke-cycle engine.
5. Explain how the valve train works.

ENGINE OPERATION

The burning of the mixture of air and fuel is called combustion. In the kind of engines we will study, combustion takes place inside the engine. For this reason, such machines are called internal combustion engines. An internal combustion engine is really just a container in which we burn air and fuel.

Basic Parts

The tube used for burning the air and fuel is called a *cylinder*, Figure 5-1. An engine cylinder is simply a metal tube closed at one end. We call the moving part that fits inside the cylinder a *piston*. There is a small space between the piston and the top of the cylinder where the burning takes place. This space is called the *combustion chamber*.

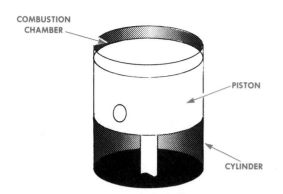

Figure 5-1. The tube is a cylinder and the plug is a piston.

As the mixture of air and fuel is burned in the combustion chamber, the expanding gas forces the piston down the cylinder. We want to use the power from the moving piston. To do this we must connect the piston to something. This can be done by attaching a rod to the bottom of the piston. The other end of the rod may be connected to a pin on the spoke of a wheel, as shown in Figure 5-2. The rod is called a *connecting rod*.

As the piston is forced downward, the connecting rod attached to the pin on the spoke of the

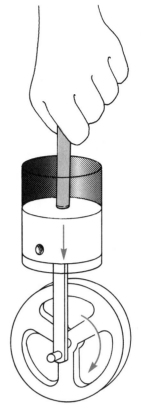

Figure 5-3. When the piston is pushed down, the wheel turns.

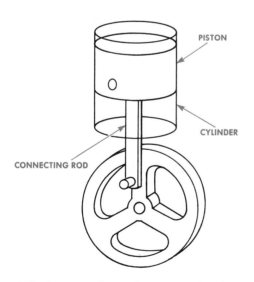

Figure 5-2. A connecting rod connects the piston to a spoke on a wheel.

wheel moves downward. This causes the wheel to turn. So a downward push on the piston is changed to a round-and-round movement at the wheel as shown in Figure 5-3.

In a real engine, we do not use a wheel with a spoke. We use a part called a *crankshaft*. The crankshaft is a bar with its ends mounted so that it can turn freely. The middle of the crankshaft is bent up or offset. The lower end of the connecting rod is connected to the middle of the crankshaft as shown in Figure 5-4. At the upper end, the connecting rod is connected to the piston with a piston pin, as shown in Figure 5-5. This lets the connecting rod follow the crankshaft's motion.

The action of the piston, connecting rod and crankshaft is similar to the action of riding a bike. When your leg pushes down on a pedal, the sprocket goes around. Your legs go up and down

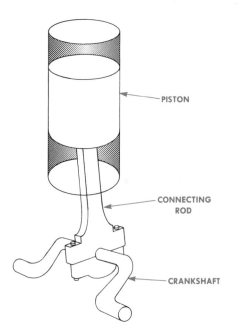

Figure 5-4. The connecting rod is connected to an offset shaft called the crankshaft.

Figure 5-6. Piston and pedal action are much the same.

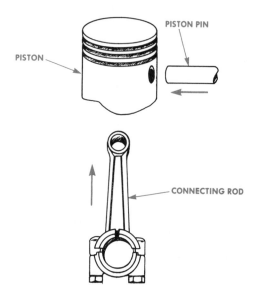

Figure 5-5. The connecting rod is attached to the piston by a pin so it can move.

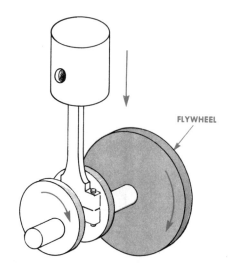

Figure 5-7. A heavy wheel called the flywheel turns with the crankshaft.

like pistons. The sprocket goes around and around like the crankshaft. This action is shown in Figure 5-6.

We need one more part to complete our basic engine. We want to push the piston down the

cylinder more than one time. This means we must bring it back up to the top of the cylinder. A heavy wheel called a flywheel is mounted to the end of the crankshaft as shown in Figure 5-7. When the piston is forced down, the crankshaft goes around. The flywheel goes around too. Since it is heavy, it does not slow down easily. The weight of the moving flywheel keeps the crankshaft turning. This movement causes the piston to go back up to the top of the cylinder.

A Stroke

In the last section, we saw how the piston was pushed down in the cylinder. The flywheel weight brought the piston back up to the top of the cylinder. When the piston shown in Figure 5-8

moves from the top of the cylinder to the bottom, we call it a *stroke*. A stroke is movement of the piston. When the piston shown in Figure 5-9 moves from the bottom of the cylinder to the top, we call that a stroke, too.

FOUR-STROKE CYCLE

In many engines, the power is developed using four piston strokes. This is why it is called a four-stroke-cycle engine. A cycle is a sequence of events that is repeated over and over.

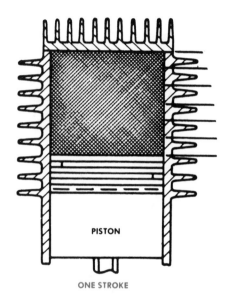

Figure 5-8. Piston movement from the top of the cylinder to the bottom is a stroke. (Briggs & Stratton Corp.)

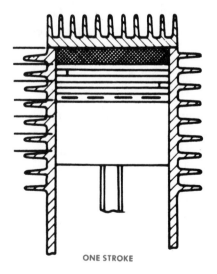

Figure 5-9. When the piston goes back up, it is another stroke. (Briggs & Stratton Corp.)

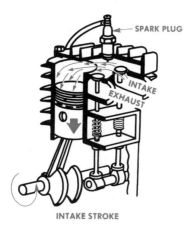

Figure 5-10. On the intake stroke the piston moves down, pulling in air and fuel. (Briggs & Stratton Corp.)

Figure 5-11. The piston moves up, squeezing the mixture for the compression stroke. (Briggs & Stratton Corp.)

Before we start, we must add something to our basic engine. We must have two holes in the top of the cylinder. These holes can be opened or closed as needed. One will be used to let air and fuel into the cylinder. It is called the intake port or passage. The other hole is used to get the burned air and fuel out of the cylinder. This is called the exhaust port or passage.

Intake Stroke

Now we are ready to see how this engine works during one complete cycle. We will start with both the intake and exhaust ports closed. The piston is as far up in the cylinder as it can be.

The first stroke is called the intake stroke, Figure 5-10. The piston moves down the cylinder very fast. This fast downward movement causes a vacuum in the cylinder. At the same time, the intake port is opened. The air-fuel mixture is pulled into the cylinder by the vacuum.

When the piston has gone down as far as it can go, the crankshaft has turned half-way around. The combustion chamber is filled with a mixture of air and fuel. We can now close the intake port.

Compression Stroke

As the piston starts back up the cylinder, the compression stroke, Figure 5-11, begins. As the piston moves to the top, it squeezes the air-fuel mixture. This squeezing of the mixture, which is called *compression,* is very important. More compression, or a tighter squeeze, results in more heat during the burning.

During the compression stroke, the crankshaft has turned another half turn. During intake and compression, it has gone completely around one time.

Power Stroke

The power stroke, Figure 5-12, starts when the piston reaches the top of the compression stroke. Here an electric spark starts the air-fuel mixture burning. The explosive force developed by the burning pushes the piston down. When the piston reaches the bottom of the cylinder, the power stroke is over. The crankshaft turns another half turn during this stroke. The exhaust and intake ports remain closed.

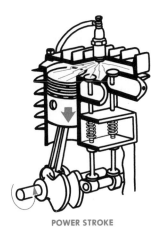

POWER STROKE

Figure 5-12. The mixture is burned, pushing the piston down for the power stroke. (Briggs & Stratton Corp.)

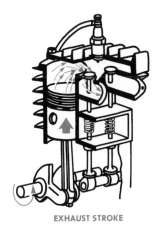

EXHAUST STROKE

Figure 5-13. The piston moves up, pushing out the burned mixture on the exhaust stroke. (Briggs & Stratton Corp.)

Exhaust Stroke

The piston starts back up the cylinder for the exhaust stroke, Figure 5-13. The exhaust port is opened. Exhaust gases are forced out of the cylinder through the exhaust port by the piston. When the piston reaches the top of this stroke, the exhaust port is closed. The crankshaft has gone around another one-half turn.

The piston can then start down for another intake stroke, and the whole cycle begins again. All four strokes are shown in Figure 5-14.

THE ENGINE'S PARTS

The engines we have been studying are called basic engines because they have only a few basic parts. A *real engine* has more parts than the basic engine. In this section, we are going to look at the parts of a real engine and see how they all fit together.

Very few of an engine's parts can be seen from the outside. There are two types of pictures that help us study parts inside an engine. One is called a cutaway. If you cut an orange down the center,

INTAKE STROKE

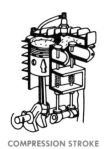

COMPRESSION STROKE

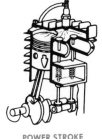

POWER STROKE

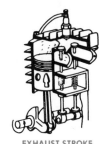

EXHAUST STROKE

Figure 5-14. The four strokes of a four-stroke-cycle engine. (Briggs & Stratton Corp.)

you can see the inside. You have a cutaway view of the inside of an orange, as shown in Figure 5-15. We can look at an engine in a similar way. A cutaway view of a four-stroke-cycle engine is shown in Figure 5-16. The outside of the engine has been cut away so the inside parts can be seen.

Another useful type of picture is called an exploded view. An exploded view shows the parts separated from each other. The parts are placed in the picture the way they would fit back together. Figure 5-17 shows an exploded view of an engine.

Figure 5-15. An orange cut this way gives us a cutaway view.

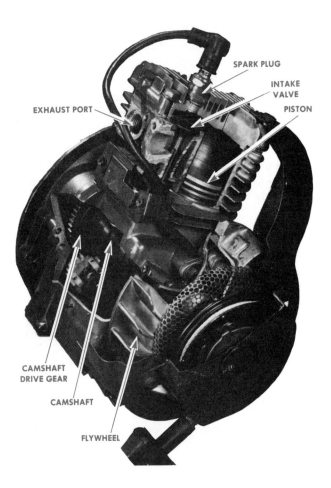

Figure 5-16. A cutaway view of an engine.

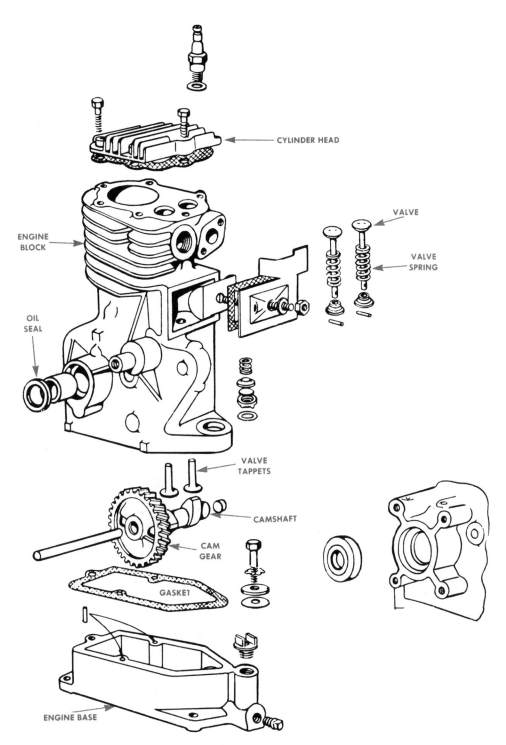

CYLINDER HEAD

VALVE

VALVE
SPRING

ENGINE
BLOCK

OIL
SEAL

VALVE
TAPPETS

CAMSHAFT

CAM
GEAR

GASKET

ENGINE BASE

Figure 5-17. An exploded view of an engine.

Crankcase and Main Bearings

Let's begin our study of the parts of an engine. The crankcase Figure 5-18, is a metal box or housing that holds the crankshaft. The crankcase may be made in one piece, or it may come apart. A hole in each side of the crankcase holds the crankshaft ends.

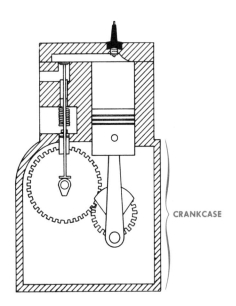

Figure 5-18. The crankcase holds the crankshaft.

MAIN BEARING

Figure 5-19. A side cover or base plate.

These holes are called the *main bearings*, or main bearing support holes. The crankshaft ends fit very closely into the main bearings, but the crankshaft can still turn. A side cover or plate is attached to the crankcase. One of the main bearings is in this plate, Figure 5-19. When the plate is taken off, the crankshaft can be removed.

Cylinder and Block

As we saw earlier, the cylinder is a hollow tube for the piston. The piston slides up and down on the cylinder walls. The cylinder must be just the right size for the piston to move freely.

Cylinders for small engines often are made from aluminum. Aluminum is a light metal which loses heat very rapidly. Aluminum cylinders are very soft. They could wear out quickly. In some engines, a thin tube of a stronger metal, either cast iron or steel, is placed in the aluminum cylinder. This is called a *liner* or *sleeve*. The piston slides on the liner instead of on the aluminum.

Cylinders on some engines are made to be removed from the crankcase. Many motorcycle

Figure 5-20. A cylinder assembly.

engines have a cylinder that can be removed. If a cylinder is damaged, it may be replaced with a new one. Most small engines have a cylinder that is made as one piece with the crankcase. When the crankcase and cylinder are one piece, the whole part is called a *block*, or cylinder assembly. A cylinder assembly is shown in Figure 5-20.

Cylinder Head and Head Gasket

The cylinder head, Figure 5-21, gives us the top for the cylinder. It is attached to the top of the block with bolts, as shown in Figure 5-22. The combustion chamber is part of the cylinder head. High pressure is built up during the power stroke.

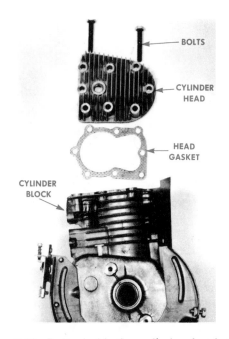

Figure 5-22. Bolts hold the cylinder head to the cylinder block.

Figure 5-21. Cylinder head.

This high pressure must not leak out between the cylinder head and block, or power will be lost. A part commonly called the *head gasket* goes between the cylinder head and block, forming a seal. The head gasket is shown in Figure 5-23.

Crankshaft

The crankshaft changes the up-and-down movement of the piston to a round-and-round movement. The parts of a crankshaft are shown in

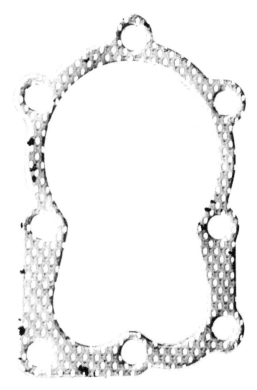

Figure 5-23. A head gasket forms a seal between the cylinder head and block.

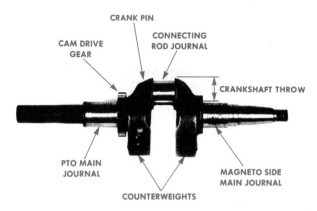

Figure 5-24. The parts of a crankshaft.

Figure 5-24. The parts of the crankshaft that fit in the main bearings are called main bearing journals. They are made very carefully for a good fit in the main bearings. The offset part of the crankshaft is called the *crankpin*. The connecting rod is attached to the connecting rod journal on the crankpin.

Some heavy weights, called counterweights, are attached to the crankshaft. These weights balance the weight of the piston and connecting rod. Counterweights help the engine run smoothly.

Crankshafts may fit into the crankcase in one of two ways. Some engines have a horizontal crankshaft. Others have the crankshaft in a vertical position. Horizontal and vertical crankshafts are shown in Figure 5-25. A horizontal crankshaft engine and a vertical crankshaft engine look different on the outside. A horizontal crankshaft engine is shown in Figure 5-26. A vertical crankshaft engine is shown in Figure 5-27.

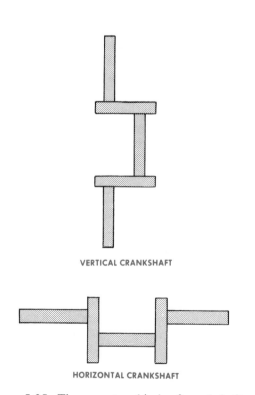

Figure 5-25. There are two kinds of crankshafts.

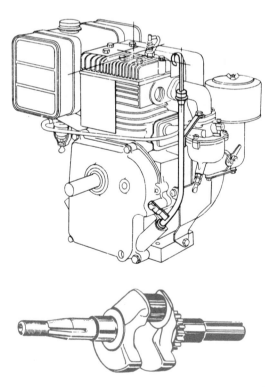

Figure 5-26. An engine with a horizontal crankshaft. (Briggs & Stratton Corp.)

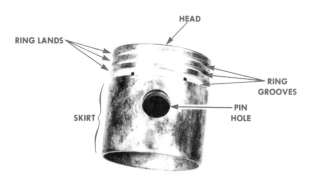

Figure 5-28. The parts of a piston.

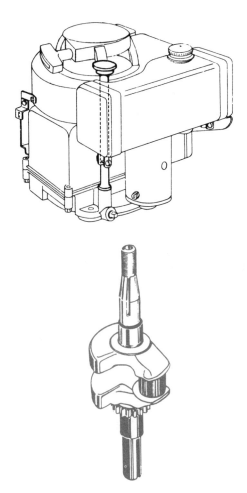

Figure 5-27. An engine with a vertical crankshaft. (Briggs & Stratton Corp.)

Piston Rings

Two types of piston rings are used, Figure 5-29. One type, called the compression ring, is located in the groove or grooves near the piston head. It seals in the compression pressure. One or more oil control rings (oil rings) are located in the groove or grooves below the compression rings. To prevent excessive oil consumption, the oil ring wipes excess oil off the cylinder wall and routes it back into the crankcase. The number and type of piston rings used are determined by the requirements of a particular engine.

Piston

The piston, which gets its push from the burning air-fuel mixture, must be strong but light. Most small engines have aluminum pistons. They are made carefully to fit in the cylinder.

The main parts of a piston are shown in Figure 5-28. The head is the top of the piston where it gets its push. Grooves are cut around the top for rings. The sides of the piston slide against the cylinder. This area is called the piston skirt. A hole in the piston, called the *pin hole,* is used to connect the piston to the connecting rod.

Figure 5-29. Piston rings seal compression and control oil.

Most small engines used two compression rings. The first compression rings made were rectangular in cross section and formed a simple mechanical seal against the cylinder wall. The rings, in their free state, were larger than the cylinder diameter. When compressed in the cylinder, they pushed out against the cylinder wall to provide a seal. Rings for modern engines make use of torsional twisting and compression pressures to improve upon a simple mechanical seal.

The job of the oil control ring is to scrape oil off the cylinder wall and direct it through the ring and into holes in the piston. Oil flows through the piston holes and runs back into the crankcase. Oil rings for small engines are one-piece rings with holes for oil flow. The pressure of an oil ring on the cylinder wall comes from tension and face width of the ring. As the face of a cast-iron oil ring wears, spring tension is decreased, and in most instances the width of the faces is increased. This results in a lower pressure and reduced oil control.

To increase ring tension, an expander, Figure 5-30, sometimes is used with an oil control ring. The expander is slightly larger around than the cylinder. When assembled behind the piston ring and in the cylinder, the expander pushes out on the ring. This forces the ring uniformly against the cylinder wall. An expander has the disadvantage of causing more rapid cylinder wear. It should be used only when recommended by the engine manufacturer.

Connecting Rod and Piston Pin

The connecting rod connects the piston to the crankshaft, Figure 5-31. The connecting rod must be strong and light. It is often made from aluminum. There is a small hole called a *pin hole* at the top of the connecting rod. A pin fits through this hole in the piston. The pin is called a *wrist pin* or *piston pin*. It may be held in place in the piston by retaining rings. The pin allows the connecting rod to move back and forth as the crankshaft turns.

The other end of the connecting rod fits around the crankshaft. To get it on the crankshaft, the rod must split apart. The rod cap is the part that comes off to let us attach the connecting rod to the crankshaft. Connecting rod bolts hold the cap to the connecting rod.

The crankshaft must turn freely when it is attached to the connecting rod. This means that

Figure 5-30. An expander often is used behind the oil control ring.

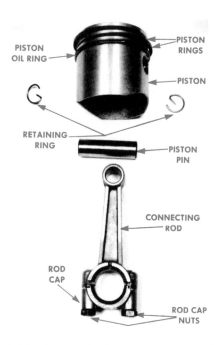

Figure 5-31. Parts of piston and connecting rod assembly.

there must be very little friction between the connecting rod and crankshaft. The rod cap must fit around the crankshaft just right. It cannot be too loose or too tight. Some engines use insert bearings between the crankshaft and connecting rod.

The Valve Train

The four-stroke-cycle engine must get the air and fuel into the cylinder on the intake stroke. We must also have a way of getting the burned air and fuel out of the cylinder on the exhaust stroke. In our basic engine, we used passageways called *ports* to get the new mixture in and the burned gases out. The ports in a real engine are opened and shut with valves. The parts used to open and shut the valves are called the *valve train*. In this section we will see how the valve train works.

Valves. The valves in an engine work much like a plug in the bottom of a sink or bathtub. When the plug or stopper is pushed into the hole, Figure 5-32, water cannot get out. If we pull the plug up out of its hole, water can get around it and down the pipe as shown in Figure 5-33.

An engine works much the same way. A valve is a round metal plug connected to a rod called a *stem*. The head of the valve is tapered, or shaped at an angle. This tapered part is called the *valve face*. When the valve is closed, the valve face seals tightly against the valve seat. The valve seat is a tapered part of the cylinder block, Figure 5-34.

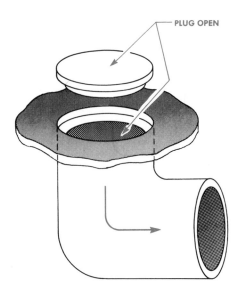

Figure 5-33. An open plug lets the water drain out.

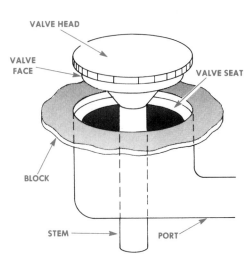

Figure 5-34. A valve is like a plug with a stem attached.

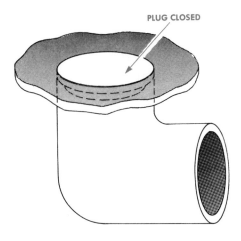

Figure 5-32. A closed plug will stop water from draining out.

There are two valves in every cylinder. One is called the *intake valve*. It is opened on the intake stroke to let air and fuel into the cylinder. The other is called the *exhaust valve*. It opens on the exhaust stroke. Burned gases can get out of the cylinder through it.

Camshaft. The job of opening the valves at just the right time belongs to the camshaft. There are

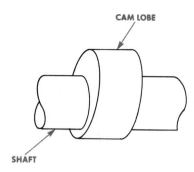

Figure 5-35. A camshaft is a shaft with a bump or lobe.

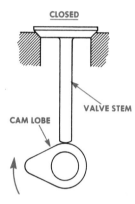

Figure 5-36. The stem is on the small part of the lobe so the valve is closed.

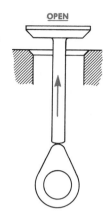

Figure 5-37. As the high part of the lobe comes around it pushes the valve open.

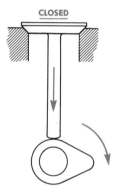

Figure 5-38. As the lobe passes by the stem, the valve closes.

bumps on the shaft called cam lobes. A camshaft with one cam lobe is shown in Figure 5-35.

The camshaft is located under the valves. When the smallest part of a cam lobe is under the valve stem, the valve is closed, Figure 5-36. As the camshaft turns, the high part of the cam lobe pushes up on the valve stem. The valve opens as shown in Figure 5-37. As the cam turns some more, the lobe passes under the stem. The valve can then be closed, as shown in Figure 5-38.

The camshaft for a small single-cylinder engine has two lobes. One lobe works the intake valve. The other works the exhaust valve.

There is a gear on the end of the camshaft. This gear fits into another gear on the crankshaft as shown in Figure 5-39. When the crankshaft turns, the camshaft is forced to turn. The camshaft gear is twice as big as the crankshaft gear, which causes the camshaft to turn only one-half as fast as the

crankshaft. They are designed this way so the valves open only on two strokes: the intake and exhaust. There are marks on the two gears, shown in Figure 5-40. These are to assist the mechanic in putting in the camshaft correctly, so the valves will open at the right times.

Valve Lifter. In most engines, the camshaft does not push directly on the valve. There are some small parts called *lifters* that ride on the cam lobes. The lifters, shown in Figure 5-41 push on the valve stems.

Valve Springs. The camshaft opens the valve. A spring, called the *valve spring*, closes the valve. The valve spring must hold the valve tight against

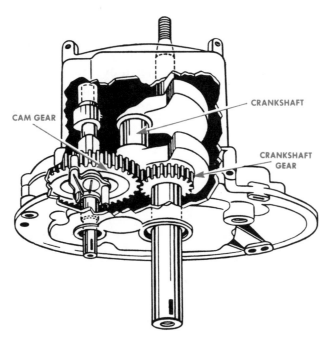

Figure 5-39. A gear on the crankshaft turns a gear on the camshaft. (Clinton Engines Corp.)

Figure 5-40. Timing marks are used to line up cam and crankshaft.

Figure 5-41. Valve lifter.

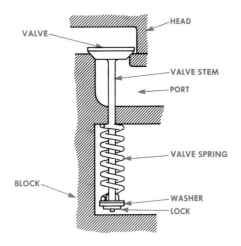

Figure 5-42. Valve spring, washer, and lock.

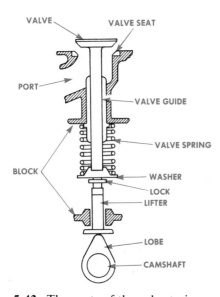

Figure 5-43. The parts of the valve train.

its seat for an airtight seal. A valve spring is shown in Figure 5-42. One end of the spring rests against the block. The valve stem goes through the spring coils. The other end of the spring rests on a small round washer. A keeper or retainer lock holds the washer on the valve stem.

When the valve is opened, the spring gets squeezed. As the cam turns around, the spring is released and pulls the valve closed again.

Valve Guide. The valve guide is a small hollow tube. It fits in the block. The valve stem fits through the guide. The guide keeps the valve centered over the valve seat.

All the parts of the valve train we have studied are shown in Figure 5-43.

NEW TERMS

camshaft: A shaft with lobes used to open the engine's valves at the proper time.

combustion chamber: Part of the engine in which the burning of the air and fuel takes place.

compression ring: A piston ring used to seal compression pressures inside the combustion chamber.

compression stroke: The stroke of the four-stroke engine during which the air-fuel mixture is compressed.

connecting rod: An engine part that connects the piston to the crankshaft.

connecting rod bearing: The device used between the connecting rod and the crankshaft to reduce friction and wear.

crankcase: The part of the engine that supports the crankshaft.

crankshaft: An offset shaft to which the piston and connecting rods are attached.

cylinder: A tube in which the piston rides.

cylinder head: Large casting bolted to the top of the engine containing the combustion chamber.

engine: A machine that converts heat energy into mechanical power to perform work.

exhaust ports: Passages used to route out burned gases from the cylinder.

exhaust stroke: The stroke of a four-stroke-cycle engine during which the burned gases are expelled.

exhaust valve: Valve used to control flow of burned exhaust gases from the cylinder.

flywheel: Heavy wheel used to store energy and smoothe out engine operation.

four-stroke-cycle engine: Engine that generates power using four strokes of a piston.

intake ports: Passages in the cylinder head used to route the flow of air and fuel into the cylinder.

intake stroke: The stroke of the four-stroke-cycle engine during which air and fuel enter the engine.

intake valve: Valve used to control the flow of air and fuel into the engine.

internal combustion engine: An engine such as the four-stroke-cycle engine in which the burning of the fuel takes place inside the engine.

lobe: A raised bump on the camshaft used to lift a valve.

piston: A round metal part attached to the connecting rod which slides up and down in the cylinder.

piston pin: A pin used to attach the piston to the connecting rod.

piston ring: Expanding sealing ring placed in a groove around the piston.

power stroke: The stroke of the four-stroke-cycle engine during which power is delivered to the crankshaft.

valve: A device for opening and closing a port.

valve guide: A part installed to support and guide the valve.

valve lifter: A part that rides on the cam and pushes open the valve.

valve spring: Coil spring used to close the valve.

valve train: An assembly of parts in an engine that opens and closes the passageways for the intake of air and fuel as well as the exhausting of burned gases.

SELF CHECK

1. Write a definition for *engine*.
2. What happens to the piston when air and fuel are burned?
3. How do the connecting rod and crankshaft change the up-and-down motion to rotary motion?
4. What does the flywheel do?
5. What fits in the main bearings?
6. What is a crankcase?
7. Where does the cylinder head fit?
8. What are the two types of piston rings?
9. Describe how a valve can open and close a port.
10. What closes the valves?

DISCUSSION TOPICS AND ACTIVITIES

1. Use a cutaway model of an engine and describe four-stroke-cycle engine operation.
2. Point out all the parts you can on a real engine.

unit 6

fundamentals of the two-stroke-cycle engine

Not all engines we use work on the four-stroke cycle. Many small engines used for mopeds, motorcycles, chainsaws, boats and snowmobiles develop power in just two strokes of the piston. Such an engine is called a two-stroke-cycle engine. Often we shorten the name and just call this engine a two-stroke or a two-cycle. Remember that a stroke is a movement of the piston from one end of the cylinder to the other. A cycle is an action that is repeated over and over. In this unit we will study the operation of the two-stroke engine.

LET'S FIND OUT: **When you finish reading and studying this unit, you should be able to:**
1. **Describe the basic parts of a two-stroke-cycle engine.**
2. **Explain how a two-stroke-cycle engine works to develop power.**
3. **Describe how a two-stroke-cycle engine differs from a four-stroke-cycle engine.**
4. **Describe the operation of a reed valve.**
5. **Name the component parts of a two-stroke-cycle engine.**

BASIC PARTS

The basic parts of a two-stroke-cycle engine are the same as those in the four-stroke-cycle with one exception: there is no valve train in a two-cycle engine. There is a cylinder and a combustion chamber. A piston in the cylinder is connected to a crankshaft by a connecting rod. A flywheel is mounted to the crankshaft.

The basic parts of a two-stroke-cycle are shown in Figure 6-1. There are two holes or ports in the cylinder. One, called the *exhaust port,* lets the burned air-fuel mixture out. The other, called the *intake port,* lets the fresh air and fuel in.

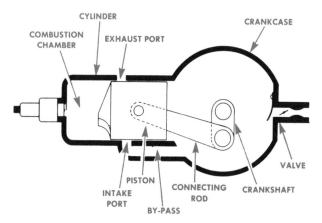

Figure 6-1. Basic parts of a two-stroke engine. (Tecumseh Products Co.)

The intake port opens into a passageway to the crankcase. This passageway may be called a bypass or transfer port. The crankcase is a box or housing that holds and protects the crankshaft. A valve in the crankcase lets in air and fuel.

HOW IT WORKS

The two-cycle engine has to do in two strokes what is done by the four-cycle engine in four strokes. When the piston moves up in the cylinder, the space below the piston gets bigger. The result is a vacuum in the crankcase. An air and fuel mixture is pulled into the crankcase through the valve, as shown in Figure 6-2. At the same time, the piston covers the intake and exhaust ports. Air and fuel above the piston are trapped. The piston squeezes or compresses the mixture in the combustion chamber, as shown in Figure 6-3.

Near the top of the piston's upward stroke we introduce a spark to start the mixture burning. The burning causes an explosion, just as it did in the four-stroke cycle. The expanding gases force the piston down. The piston gives its power to the crankshaft, as shown in Figure 6-4.

As the piston moves down, the crankcase area becomes smaller and the air and fuel mixture in the crankcase is squeezed. The mixture would like to escape, but the valve is closed. The mixture gets squeezed more and more tightly in the crankcase.

Finally, the piston goes far enough down the cylinder to uncover the exhaust port. The burned

mixture goes out of the cylinder through the exhaust port, as shown in Figure 6-5. When the piston reaches the bottom of its stroke, it uncovers the intake port. Air and fuel trapped in the

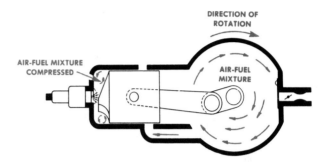

Figure 6-3. The piston squeezes the mixture in the combustion chamber. (Tecumseh Products Co.)

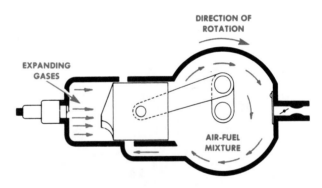

Figure 6-4. The mixture is burned, pushing the piston down. (Tecumseh Products Co.)

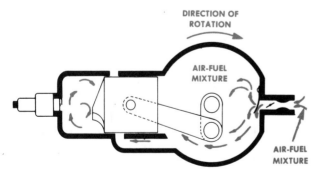

Figure 6-2. The piston moves up, pulling the mixture into the crankcase. (Tecumseh Products Co.)

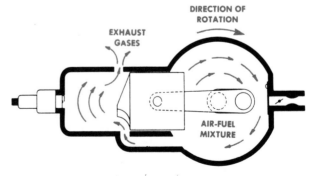

Figure 6-5. As the piston uncovers the ports, burned gases escape and a new mixture gets in. (Tecumseh Products Co.)

crankcase flow through the bypass or transfer port and into the cylinder. The piston moves back up, trapping the mixture, and the cycle starts all over again.

The two-stroke-cycle engine develops power in two strokes. On one stroke the piston compresses the mixture. On the second stroke it is pushed down the cylinder. The crankshaft turns one complete turn or revolution during these two strokes.

TWO-STROKE-CYCLE ENGINE PARTS

A cutaway view of a two-stroke-cycle engine is shown in Figure 6-6. The two-stroke engine has many parts which are similar to the four-stroke cycle. It does not, of course, have any valve train. In this section we will study the parts which are different from the four-stroke-cycle engine.

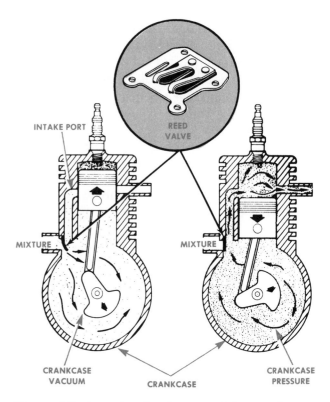

Figure 6-7. A reed valve is opened by crankcase vacuum and closed by crankcase pressure. (McCulloch Corp.)

Figure 6-6. Cutaway of a two-stroke-cycle engine.

Reed Valve

A valve is used to let the mixture into the crankcase when the piston moves up the cylinder. The valve must close when the piston moves down, so the mixture can be trapped and squeezed. Many small engines use a type of valve called a *reed valve*. The reed valve is a thin piece of metal. It works much like a hinge. When there is a vacuum in the crankcase, the reed valve is pulled

open, as shown in Figure 6-7, *left*. This allows the mixture to enter the crankcase.

When the piston moves down, the area of the crankcase gets smaller. Pressure starts to build up in the crankcase. The pressure pushes the hinged valve closed, as shown in Figure 6-7, *right*, and the mixture is trapped in the crankcase.

There are many types of reed valves. The valves are attached to a metal part called a *reed plate*. A reed valve is shown in Figure 6-8.

Crankcase and Main Bearings

The crankcase for a two-stroke engine is shown in Figure 6-9. The crankcase houses and supports the crankshaft. The crankcase is also the area into which the air-fuel mixture enters when the piston moves up. A mount for the reed valve may be attached to the crankcase.

Figure 6-8. A reed plate with four reed valves.

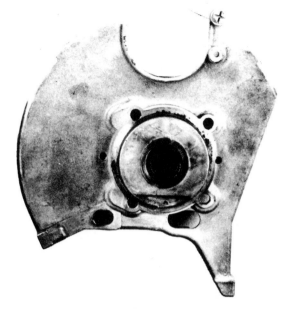

Figure 6-10. Ball or needle bearings may be used for two-stroke main bearings.

The crankcase provides the support for the main bearings. Two-stroke engines often use *ball* or *needle* bearings for main bearings, as shown in Figure 6-10. Ball and needle bearings allow the engine to run at a higher speed.

Cylinder and Block

Two-stroke cylinders generally are made from aluminum with a cast-iron liner. The aluminum makes the engine light in weight and helps get rid of heat quickly. The cast-iron liner provides a good bearing surface for the piston rings.

Figure 6-9. Crankcase for a two-stroke-cycle engine.

Figure 6-11. Two-stroke cylinders have ports for intake and exhaust gases.

The main difference between a two- and four-stroke cylinder are the intake and exhaust ports. The two-stroke has holes or slots cut directly into the cylinder as shown in Figure 6-11. These are uncovered by the piston to allow air and fuel in and burned gases out.

Cylinder Head

Some two-stroke engines have a removable cylinder head just like a four-stroke engine. The cylinder head is attached with bolts or screws. A gasket fits between the cylinder and cylinder head to form a seal. The cylinder head forms the combustion chamber.

Some two-stroke engines do not have a removable cylinder head. The head and combustion chamber are cast as one piece with the cylinder. This makes the engine lighter and eliminates any problems from leaking cylinder head gaskets. A one-piece cylinder and cylinder head is shown in Figure 6-12.

Figure 6-12. A one-piece cylinder and cylinder head is common in two-stroke engines.

Crankshaft

The two-stroke-cycle crankshaft has the same job as the crankshaft in a four-stroke-cycle engine. It changes the up-and-down movement of the piston to round-and-round movement. The

Figure 6-13. A two-stroke-cycle engine crankshaft.

two-cycle crankshaft, shown in Figure 6-13, has the same parts as a four-stroke crankshaft.

Piston

The pistons used in two-stroke engines are very different from those used in four-stroke engines. The top of the piston often has a high dome or deflector, as shown in Figure 6-14. The deflector prevents the burned gases from mixing with the fresh charge of air and fuel. When the piston uncovers the intake and exhaust ports, the deflector acts as a wall to help prevent the two gases from mixing.

Figure 6-14. A piston with a deflector, or contoured top, helps prevent mixing of intake and exhaust gases.

The ring area of the piston is also different from that of a four-cycle piston. There are often only two ring grooves. Sometimes, the ring grooves have a pin or stop mounted in them, as shown in Figure 6-15. This prevents the piston

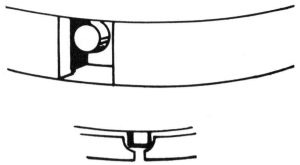

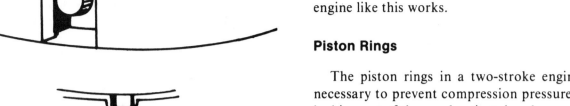

Figure 6-15. A stop pin in the ring groove prevents the ring from rotating.

lined up, the air-fuel mixture may be routed through the piston. We will see later how an engine like this works.

Piston Rings

The piston rings in a two-stroke engine are necessary to prevent compression pressure from leaking out of the combustion chamber area. Oil control is not as important in this engine as it is in a four-stroke. We will see why when we study the operation of the lubrication system in a later chapter. The two-stroke engine generally uses just two rings, both compression rings, Figure 6-17.

Figure 6-16. Some pistons have holes or slots in the skirts for mixture flow.

Figure 6-17. A two-stroke engine uses two compression sealing rings.

ring from rotating around the ring groove. The ends of the ring must not be allowed to rotate around into the port area. If they do, the ends of the ring could catch in the ports and the ring could break.

Some two-stroke pistons have holes or slots in the skirt area, Figure 6-16. These slots or holes line up with passages in the cylinder and crankcase when the engine is running. When they are

Connecting Rod and Piston Pin

The connecting rod and piston pin are similar in design to those used in a four-stroke engine. The connecting rods, Figure 6-18, often are made of cast iron or steel to provide needed strength and to permit the use of needle bearings. Needle bearings often are used in both ends of the connecting rod. They allow for high-speed operation and can be lubricated easily. The piston pin usually is held in the piston with two lock rings similar to those used in the four-stroke engine.

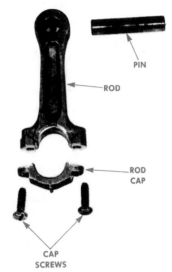

Figure 6-18. A connecting rod used in a two-stroke engine.

Loop-Scavenged Two-Stroke Engines

As mentioned earlier some two-stroke engine pistons have holes or slots in the skirt area. Such pistons are used in machines described as loop-scavenged engines. A loop scavenged engine works just like the two-stroke engine described earlier. The only difference is that the air-fuel mixture travels or loops through the holes in the pistons.

The operation of a loop-scavenged engine is shown in Figure 6-19. The piston is constructed with large bores in the piston skirt. When the piston is at the bottom of its stroke, these bores line up with passages in the cylinder. The fuel-air mixture, compressed slightly during the power stroke of the piston, passes through the bores and mixes with the fuel-air mixture at two sides of the cylinder. This eliminates the need for the conventional piston with a deflector. The loop-scavenged type of engine produces a somewhat greater horsepower per unit weight by more completely removing the exhaust gases at the end of the power stroke.

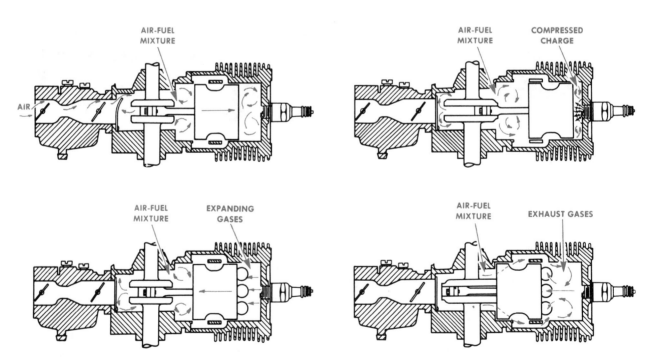

Figure 6-19. Operation of a loop-scavenged two-stroke-cycle engine. (Tecumseh Products Co.)

NEW TERMS

bypass port: Passageway between the crankcase and combustion area in a two-stroke engine

loop-scavenged: A type of two-stroke engine in which the air-fuel mixture loops through bores in the piston skirt.

reed valve: A valve used to control the flow of air-fuel mixture into the crankcase.

rotary valve: A valve attached to the crankshaft that controls the flow of air and fuel into the crankcase.

transfer port: Same as bypass port.

two-stroke-cycle engine: An engine that develops power in two piston strokes or one crankshaft revolution.

SELF CHECK

1. How many strokes does it take to develop power in a two-cycle engine?
2. What is another name for a two-cycle engine?
3. What are the names of the two ports in a two-cycle cylinder?
4. Describe what happens when the piston moves up in a two-cycle engine.
5. Describe what happens when the piston moves down in a two-cycle engine.
6. What is the purpose of the bypass or transfer port?
7. How does the burned mixture get out of the cylinder?
8. How does the fresh mixture get into the cylinder?
9. How does a reed valve work?
10. Why is a liner used in the aluminum cylinder?
11. Why does a two-cycle piston have a deflector?
12. Why does a two-cycle engine not use an oil control ring?
13. Why does a loop-scavenged engine have holes in the piston?

DISCUSSION TOPICS

1. Use a cutaway model to identify the major parts of a two-stroke-cycle engine.
2. Turn the crankshaft of the cutaway model one revolution and explain what happens during each stroke.

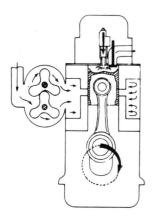

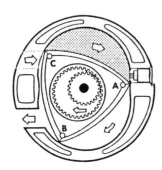

unit 7

diesel and rotary valve engines

Two additional types of small engines are in use. One is the diesel engine. This engine, long popular in trucks, is becoming more popular for small-engine use. Small diesel engines are used to power generators and water pumps. The diesel uses the heat produced by high compression to ignite the air-fuel mixture.

In recent years, another type of internal combustion engine has been used in motorcycles, outboards and snowmobiles. This engine is known as the rotary, rotary combustion or Wankel, after its inventor, Felix Wankel. The rotary engine performs the same job as a four-stroke-cycle engine. Its working parts, however, have a rotary rather than an up-and-down, or reciprocating, motion. In this unit we will see how these two engines work.

LET'S FIND OUT: **When you finish reading and studying this unit, you should be able to:**
1. **Explain how the air-fuel mixture is ignited in a diesel engine.**
2. **Describe the operation of a two-stroke-cycle diesel engine.**
3. **Describe the operation of a four-stroke-cycle diesel engine.**
4. **List the basic parts of a rotary engine.**
5. **Explain the operation of a rotary engine.**

DIESEL ENGINE OPERATION

The diesel engine, Figure 7-1, has the same basic components as other engines we have studied. The main component of the diesel engine is the cylinder block which contains a cylinder or number of cylinders in which the pistons move up and down. The piston compresses the charge of air in the cylinder and receives the pressure of the expanding gases during combustion.

The connecting rod, connected to the piston by the piston pin, transmits the force on the piston to the crankshaft. The crankshaft converts the reciprocating piston motion to rotary motion. The flywheel on the end of the crankshaft maintains the smooth rotary motion.

A cylinder head closes off the top of the cylinder to confine the fuel and air. Valves in the cylinder head admit air into the cylinders and discharge the exhaust gases. The engine camshaft opens and closes the valves at the right time.

Diesel engines can operate on either the two-stroke or the four-stroke cycle. In the two-stroke engine, Figure 7-2, an air pump, called a *supercharger,* blows air into the cylinder through intake ports located near the bottom of the cylinder. Fuel is injected at the top of the compression stroke through a fuel injector and ignites upon contact with the compressed air. The rapid expansion of the burning fuel drives the piston

Figure 7-1. A small diesel engine.

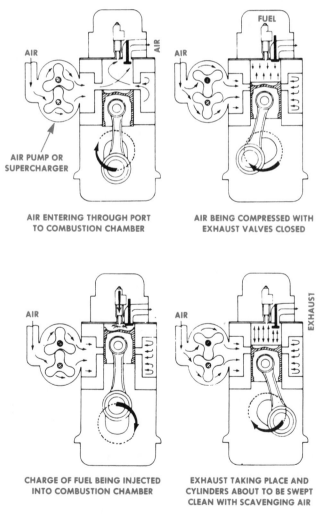

Figure 7-2. Two-stroke diesel operation. (Go Power)

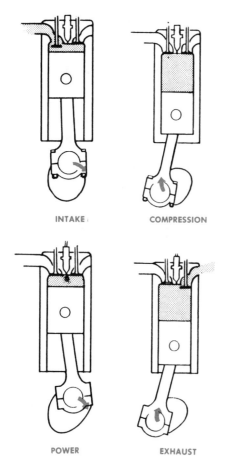

INTAKE   COMPRESSION

POWER   EXHAUST

Figure 7-3. Operation of a four-stroke-cycle diesel. (Onan Corp.)

down against the crankshaft for the power stroke. Near the end of the power stroke, the exhaust valve opens to permit exhaust gases to escape. A short time later, the air intake ports are again uncovered and fresh air is forced into the cylinder, pushing out any remaining exhaust gases and filling the cylinder for the next compression stroke.

In a four-stroke-cycle diesel engine, Figure 7-3, the order of events is the same as in the four-stroke-cycle spark-ignition engine.

During the intake stroke, the piston moves down rapidly, causing a vacuum in the cylinder. Air, under atmospheric pressure, rushes into the cylinder through the intake valve to fill the vacuum.

When the piston reaches the bottom of the cylinder, the intake valve closes. As the piston moves upward for the compression stroke, the air is compressed to a much greater extent than in a spark-ignition engine. At highest compression, the pressure is so great that the air reaches temperatures of 800 to 1100 degrees Fahrenheit (427 to 594 degrees Celsius).

When the piston reaches the top of the cylinder, the fuel valve opens and fuel oil is sprayed under pressure into the superheated air in the combustion chamber. At this temperature, smaller drops of fuel start burning immediately. This increases pressure and temperature, causing larger and larger particles to burn. Combustion does not happen all at once in a diesel engine. It is spread out in a wave or front. This creates a gradual increase in pressure within the cylinder, driving the piston downward for the power stroke.

When the piston reaches the end of the power stroke, the exhaust valve opens and the piston moves upward for the exhaust stroke, forcing exhaust gases out.

DIESEL ENGINE PARTS

As mentioned previously, most of the diesel engine parts are the same as those in other small engines. The main difference is strength. In most of today's gasoline engines, compression ratios range from about 8:1 to 10:1. Diesel engine compression ratios are nearly all from around 10:1 to 20:1. The higher compression ratio results in increased combustion pressure on the power stroke. The much higher forces developed in the high-compression combustion chamber require that the diesel engine be much heavier than a gasoline engine.

The cylinder block in a small diesel engine supports the operating parts just as in a gasoline engine. Diesel engine blocks usually are made of cast iron. Cast iron provides the necessary rigidity. The blocks contain oil passages to lubricate the engine. Cooling passages in the block circulate coolant around the cylinder and heads to carry off the wasted heat of combustion.

Because of the high-speed rubbing action of the piston and rings in the cylinder, the cylinder bore

Figure 7-4. A wet liner for a diesel engine cylinder.

Figure 7-5. A dry liner or sleeve may be used in a diesel engine.

needs a highly wear-resistant wall. Many cylinder bores contain liners or sleeves. The cast alloy-steel liners are able to withstand high combustion temperatures and transfer heat readily to the surrounding water jacket.

Most liners are flanged at the top, and the flange fits tightly into a counterbore in the block. A cooper gasket or a very close fit seals the liner to the bore so that pressure won't be lost. The liner is held firmly in place by the pressure of the bolted-on cylinder head.

Liners may be either the wet or dry type. A wet liner, Figure 7-4, forms the inside wall of the water jacket so that heat can be transferred directly from the cylinder to the coolant. A dry liner, Figure 7-5, is in contact with the cylinder block casting, which forms the water jacket wall. A dry liner must fit very tightly in the cylinder for effective heat transfer. Wet liners are sealed on the outside by O-rings or similar seals above and below the water jacket. This keeps engine lubricant from getting in the coolant.

Cylinder heads must be very strong to contain

SKIRT THICKNESS

Figure 7-6. A diesel engine piston must have a thick skirt and head to withstand high compression pressure.

the great force developed in the combustion chambers. They must also be able to withstand large amounts of heat. Large internal passages are necessary to provide maximum coolant circulation.

The piston, Figure 7-6, must be thick enough to take the forces from combustion and to provide a good path for heat to flow from the top to the

Figure 7-7. Diesel connecting rods must be strong to transmit the power developed under high compression.

Figure 7-8. A crankshaft from a diesel engine.

bustion power from the reciprocating piston and converts it to rotating motion.

Crankshafts are very heavily built to resist bending when the cylinder fires. The crankshaft is supported by two strong main bearings, similar in design to the connecting rod bearings. Oil passages in the cylinder block and in the shaft supply lubrication to the bearings and journals.

ROTARY ENGINE OPERATION

The operation of the rotary engine is very different from that of the internal combustion piston engine. The piston engine uses pistons moving up and down in cylinders that are connected by connecting rods to the crankshaft. The connecting rods and crankshaft change the reciprocating motion of the pistons to rotary motion. The rotary engine, on the other hand, has no pistons. Instead, it uses a triangle-shaped rotor attached to an eccentric cam on a shaft called an *output shaft*.

The rotor is positioned inside an oval-shaped firing chamber, the chamber in which combustion takes place, as shown in Figure 7-9. The rotor, with seals on each of its three points, is designed so that it can rotate in the oval housing or firing chamber. Both ends of the firing chamber are

piston rings. To compensate for heat expansion, the top is tapered slightly. The piston's diameter is largest where the top meets the skirt or lower part of the piston. Often the dome of the piston is recessed to form part of the combustion chamber. Piston rings are fitted into grooves near the top of the piston. In diesel engines, there may be as many as three to six compression rings and one or more oil rings per piston.

The connecting rod, Figure 7-7, must be very strong to transmit the power developed in the cylinder to the crankshaft. Connecting rods are made from alloy steel with "I" or channel cross-sections for rigidity.

The crankshaft, Figure 7-8, receives the com-

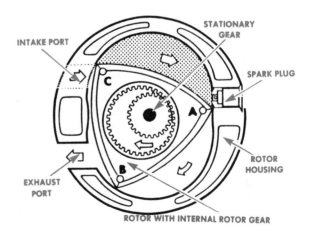

Figure 7-9. The rotor fits inside an oval firing chamber.

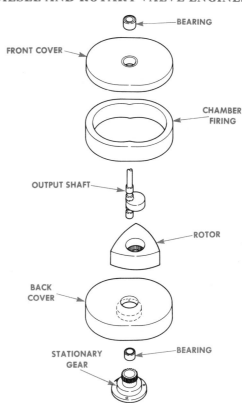

Figure 7-10. Basic rotary engine parts. (Go-Power Corp.)

closed and sealed. Two passageways, or ports, are located in opposite sides of the chamber. One port is connected to a fuel system to allow air and fuel to enter the engine. The other is connected to the vehicle exhaust system, to collect burned gases. A spark plug is screwed into the chamber.

The rotor is offset from the output shaft by the eccentric cam. As it rotates, the ends of the rotor go around in the elliptical chamber. A stationary gear, centered on the output shaft, engages teeth on an internal gear attached to the rotor. The stationary gear does not rotate, but causes the rotor to follow the correct path within the firing chamber, see Figure 7-10. The gear ratio between the internal rotor gear and the stationary gear is 3 to 1. This ratio causes the output shaft to rotate three times for every rotation of the rotor.

As one face of the triangular rotor sweeps past

the intake port, Figure 7-11, (1-4) (see key at bottom of figure) a partial vacuum occurs in this part of the firing chamber. Air and fuel are forced through the intake port into the firing chamber. As the rotor continues to turn, the seal crosses the intake port, sealing off the chamber. The rotor continues to turn until the air-fuel mixture is compressed into the smallest volume, Figure 7-11 (5-9). The spark occurs at maximum compression. Rapidly expanding gases pushing against the rotor face force the rotor to continue turning in a clockwise direction, Figure 7-11 (10-12). Power on the face of the rotor is transmitted over the eccentric to the output shaft. As the apex clears the exhaust port, the exhaust gases are swept out of the chamber. Exhaust is completed as the apex sweeps past the exhaust port, Figure 7-11 (13-18).

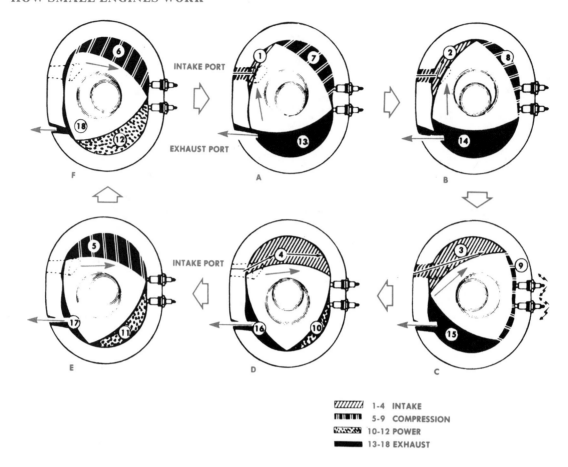

INTAKE PORT

EXHAUST PORT

INTAKE PORT

	1-4 INTAKE
	5-9 COMPRESSION
	10-12 POWER
	13-18 EXHAUST

Figure 7-11. The operation of a rotary engine.

ROTARY ENGINE PARTS

A rotary engine may be equipped with one, two or even three or more rotors. Typically, a small rotary engine has a single rotor. When more than one rotor is used, they are positioned out of phase with each other, to provide an overlapping of power much like multiple cylinders. In this section we will examine each component of a single-rotor engine.

Rotor Housing

The rotor housings form the firing chamber for the rotor. The rotor housings perform the same function as the cylinder block in the piston engine. The housings are stacked together, as shown in Figure 7-12. A one-rotor rotary engine may have three housings: a front, a rear and a middle, or rotor, housing. The middle housing provides the firing chamber for the rotor. The front and rear housings seal the ends of the firing chamber and support the output shaft.

The middle housing provides the working surface for the rotor. This housing may be made of aluminum or cast iron, but the working surface must be extremely hard and durable to withstand the constant action of the rotor seals. Aluminum, when used, is usually chrome-plated. An exhaust port is machined into the housing. An exhaust manifold bolted to the outside of the housing directs exhaust gases into the exhaust system. The firing area of the chamber has a hole for installation of a spark plug.

Figure 7-12. A single-rotor engine has two housings.

Output Shaft

The output shaft for a single-rotor engine is shown in Figure 7-13. The shaft may be cast or forged from cast iron or steel. It is machined to very close tolerances. Two main journals are machined on either end. These fit into bearings in the front and rear housings. An eccentric-shaped rotor journal is provided for the rotor. These journals are offset so that the rotor may apply a rotational force to the output shaft in much the same way the piston and connecting rod in a piston engine push on the crankshaft.

Rotor

The rotor has the same function as the pistons in a four-stroke-cycle engine. It receives the force from combustion and transfers this force to the output shaft. The rotor is positioned in the rotor housing. The parts of a rotor are identified in Figure 7-14. Each of the rotor's three sides has an indention that serves as a combustion chamber. A rotor bearing is pressed into the center of the rotor. The bearing provides the running surface for the eccentric rotor journals on the output shaft. Internal gear teeth are machined into the center of the rotor. This gear meshes with a stationary gear mounted to the front and rear housing. The meshing of the two gears insures that the rotor will follow the correct path as it rotates in the housing.

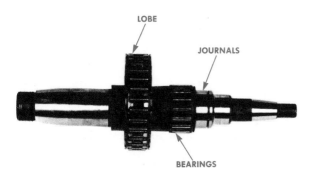

Figure 7-13. Parts of a rotary engine output shaft.

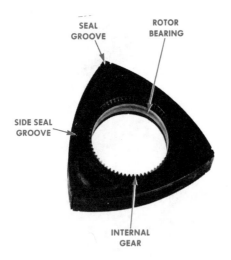

Figure 7-14. Parts of the rotor.

Rotor Seals

The rotor illustration in Figure 7-15 shows the three types of seals necessary to seal the combustion pressures in the rotor housing. These seals have the same job as piston rings in a piston engine. Side seals fit in a groove on each side of the rotor. The side seals prevent compression pressure from leaking between the side of the rotor and the face of the housings.

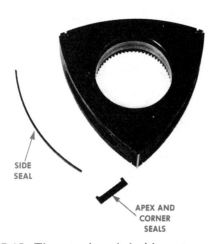

Figure 7-15. The rotor is sealed with apex, corner, and side seals. The corner and apex seals may be separate (three pieces) or one piece as shown here.

The apexes of the rotor provide the most difficult sealing job. All three sides of the rotor must be sealed from each other so that intake, combustion, and exhaust can occur. This requires an apex seal that can maintain contact with the rotor housing as the rotor rotates, sealing each end. The difficult job of sealing the apexes of the rotor was one of the biggest problems in developing a practical rotary engine. An apex seal is mounted in a groove machined in the apex area of the rotor. A spring positioned behind the seal holds it in contact with the rotor housing. A corner seal prevents leakage at the junction of the side and apex seals.

NEW TERMS

apex seals: Seals on the ends of apexes of the rotary engine rotor, used to seal the firing chamber.

diesel engine: An engine that uses the heat of compression to ignite the air-fuel mixture.

firing chamber: The area in the rotary engine where power is developed.

output shaft: A shaft in a rotary engine that is driven by the rotor. The power to drive a vehicle's wheels comes through this shaft.

rotary engine: An engine that develops power using the explosive power of burning gasoline to push against a triangle-shaped rotor.

rotor: The triangle-shaped part in a rotary engine that has the same job as a piston in a four-stroke-cycle engine.

rotor seals: Apex, corner and side seals used to prevent leakage out of the firing chamber.

side seals: Seals installed in the side of a rotor to seal off the firing chamber.

Wankel engine: Same as rotary engine. Named after the inventor, Felix Wankel.

SELF CHECK

1. How is the air-fuel mixture ignited in a diesel engine?
2. Explain how a two-stroke-cycle diesel engine works?
3. Explain how a four-stroke-cycle diesel engine works?

4. Why must diesel engine parts be stronger than gasoline engine parts?
5. Explain the operation of the rotary engine by comparing the action of the rotor to the pistons in a piston engine.
6. What part in the rotary engine takes the place of pistons in a piston engine?
7. What is the firing chamber?
8. What does the output shaft do?
9. What part of the rotary engine does the same job as the piston rings on a piston engine?
10. Explain the purpose of the apex seals on a rotor.

DISCUSSION TOPICS AND ACTIVITIES

1. Use a cutaway engine or model to identify the major parts of a diesel engine. Explain how it works.
2. Use a cutaway or model of the rotary engine to identify the major parts. Can you explain the operation of the engine?

Snow trikes use small engines.
(Heald)

part 3
small engine systems

Regardless of the type of small engine, a number of systems are necessary to make it work. A lubrication system is needed to reduce friction and prevent engine wear. A cooling system is required to regulate the engine's temperature within safe limits. To operate, the engine must be provided with the correct amount of air and fuel. The mixture of air and fuel must be ignited inside the cylinder at just the right time. A fuel system and an ignition system handle these jobs.

The purpose of the ignition system is to provide a high-voltage spark in each of the engine's cylinders at the right time to cause the air-fuel mixture to burn. The system must develop the 25,000 or more volts required for ignition. The high voltage must be distributed to the cylinder just as it is ready for a power stroke.

The engine runs on a mixture of air and fuel. The fuel system has three functions: (1) to provide a way of storing enough fuel for several hours of operation; (2) to deliver the fuel to the engine and (3) to mix the fuel with air in the proper amounts for efficient burning in the cylinder.

The lubrication system performs a number of important functions in the engine. The system is designed to circulate oil between moving parts to prevent metal-to-metal contact and the wear which would result. Oil between moving parts allows them to move more easily, with less friction. The lower the internal friction of an engine, the more power it can develop. The circulating oil helps in cooling the engine, as it carries heat away from hot engine components. It also cleans or flushes dirt and deposits from the engine parts. Finally, the oil which is circulated on the cylinder walls helps the rings to seal tightly and thereby improves the engine's compression.

The burning of the air-fuel mixture that occurs during the power stroke of an engine generates tremendous heat. A cooling system is required to remove some of the heat in order to have an efficient engine operating temperature. The cooling may be done with air or water.

71

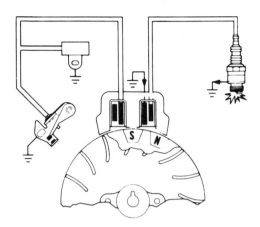

unit 8

ignition systems

An engine develops power when the heat of a burning air-fuel mixture pushes the piston down. An electric spark starts the air-fuel mixture burning. The spark comes from a system on the engine called the *magneto*.

A magneto needs several parts to make a spark. The engine in Figure 8-1 shows the basic parts of a magneto: a magnet, armature, coil, spark plug, breaker points, and condenser. In this unit we will study how a magneto ignition system works.

LET'S FIND OUT   When you finish reading and studying this unit, you should be able to:
1. Define the terms *electricity* and *magnetism*.
2. Identify the parts of a magneto ignition system.
3. Describe the operation of the magento breaker points and condenser.
4. Explain the operation of the magneto magnets and coil.
5. Describe the parts and operation of a spark plug.

ELECTRICITY

In order to understand how a magneto works, you need to know a little about electricity and magnetism. The current theory or way of thinking about electricity is called *the electron theory*. We still do not know everything about electricity. Using the electron theory we can, however, understand how electricity behaves and how to use it. In order to understand the electron theory, it is necessary to look briefly at what the scientists call the composition of matter.

Everything in the universe except the complete voids that exist between the sun, stars and planets is called matter. Anything that has weight and takes up space is matter. Even things that cannot be seen, such as air, are matter. Matter may be in the form of a solid, a liquid or a gas. All matter is

72

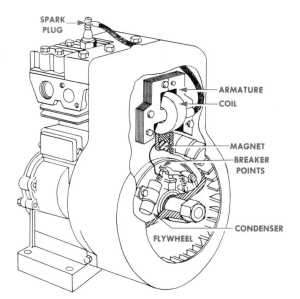

Figure 8-1. The basic parts of a magneto. (Briggs & Stratton Corp.)

composed of very, very small particles called *atoms*. An atom is so small that it is not visible except under the most powerful electron microscope. Atoms are made up of even smaller particles.

An atom is constructed much like our solar system, Figure 8-2. Think of the sun and the various planets which revolve around it. An atom has a center or core composed of particles called *protons* and *neutrons*. This core is called a *nucleus*. In our example of the solar system, the sun is like the nucleus. Other small particles called *electrons* circle in orbits around the nucleus, much as the planets circle the sun. Electrons travel at a tremendous rate of speed.

The particles which make up the atom have positive and negative electrical charges. Why atoms have this quality is not known but this is where electricity comes from. *Positive and negative charges* means simply that the two charges are completely opposite. The symbol + is used to show a positively charged particle and the symbol – to show one with a negative charge. Almost everyone has experimented with a set of magnets and observed that the magnets can be placed so that they repel each other or so that they attract each other. Electrical charges act in much the same way. Two particles that are positively charged will repel each other; two negatively charged particles will repel each other. A positively charged particle and a negatively charged particle attract each other. This positive-negative attraction is what holds the atoms together.

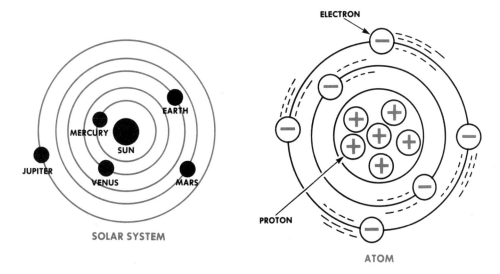

Figure 8-2. An atom is constructed like our solar system.

The core, or nucleus, of an atom is made up of positively charged particles. The electrons that orbit in a fixed pattern around the nucleus are negatively charged particles. The only difference among the atoms of different kinds of matter is the number of particles in the nucleus and the number and spacing of the electrons that orbit the nucleus.

The nucleus of the atom is composed of protons with a positive charge and neutrons with a neutral charge. The electrons with a negative charge orbit a specific distance away from the nucleus. An atom may have one, two or three rings of electrons depending on the number of electrons it contains. Each of these rings requires a specific number of electrons.

The Flow of Electricity

If we again compare the atom to our solar system, the electrons orbit the nucleus as the planets orbit the sun. The electrons remain in their orbit around the nucleus because of the electrical attraction the electrons have for the nucleus. This is similiar to the gravitational pull of the sun on the earth. The electrons that orbit closest to the nucleus are strongly attracted to it. These are called *bound* electrons. The electrons that are farther away from the pull of the nucleus can be forced out of their orbits. These are called *free* electrons. Free electrons can move from one atom to another. The movement of the free electrons from one atom to another is known as *electron flow.*

According to the electron theory, electricity is the movement or flow of electrons from one atom to another, Figure 8-3. In order to have a movement of electrons, it is necessary to have a condition of imbalance. In a normal atom, the positively charged nucleus balances the negatively charged electrons and holds them in orbit. If an atom loses electrons, it will become positive in charge. It will attract more electrons in order to regain its balance.

The flow of electricity is made possible by causing electrons to leave their atoms and gather in a certain area, leaving behind atoms without their normal number of electrons. Science has discovered a number of ways to create an unbalanced condition to start an electron flow.

Language of Electricity

Electricity has a language all its own. A mechanic must understand the meaning and relationship of a number of electrical terms in order to understand magnetos. The terms described below are relatively simple but they are the foundation for all electrical troubleshooting and servicing.

Voltage (volt) (E). In order to have a flow of water in a fire hose, pressure is necessary. In order to have an electron flow in an electrical system, pressure is necessary. In electricity the force or potential that pushes electrons is called *voltage.*

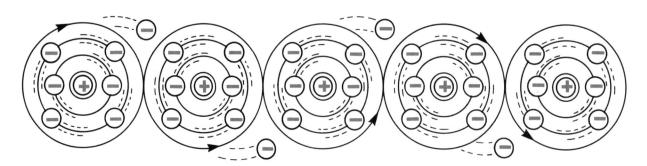

Figure 8-3. Electricity is the movement of electrons from one atom to another.

Water pressure is measured in pounds per square inch; electrical pressure is measured in volts. The letter symbol for the volt is E (for electromotive force) or V.

Voltage may be considered a source of potential energy that exists when unequal numbers of electrons are present in a system. *Voltage* or *volts* always describes a potential difference between two parts of an electrical system. The voltage of your house wiring may be 110. This voltage is present even though no household appliance is turned on. The voltage "stands by" until an appliance is turned on. It is important to understand that voltage can exist without electron flow although electron flow cannot exist without voltage.

Current (ampere) (I). Returning to the example of the fire hose, the rate of water flow may be measured in gallons per minute. In an electrical system, we are interested in the rate of electron flow. The flow of electrons is called *current*. Current is measured in amperes. The letter symbol for current is I. The flow of current is measured by a specific number of electrons passing a given point in one second. One ampere is equal to 6.28 billion electrons per second. As was mentioned previously, current cannot flow unless there is pressure or voltage.

Resistance (ohm) (R). The diameter of the fire hose will determine the amount of water that will be able to flow through it in a given amount of time. A smaller hose will provide more resistance to the flow. There is also a resistance to electron flow in an electrical system. Resistance is the opposition offered by a material to the free flow of electrons. The unit of resistance is called an ohm. The letter symbol is R.

When current runs into resistance, two things occur: first, the electrons must work harder to get through, and this creates heat. Second, the rate of their flow is reduced because some of the energy is used up as heat. The heat built up by resistance is sometimes used to do work. For example, in an ordinary household toaster, current is directed through a strong resistance. The heat produced in the resistance is used to toast the bread.

Conductor. A conductor is any material that allows a good electron flow. In a small-engine electrical system, copper and aluminum wires are used to conduct electricity because they allow good electron flow. To be a good conductor, a material must be made of atoms that give off free electrons easily. Also, the atoms must be close enough to each other so that their free-electron orbits overlap. Of all the metals, silver is the best conductor but it is too expensive for general electrical use.

Insulator. Insulators are materials whose atoms will not part with any of their free electrons. These materials will not conduct current. The copper wire in a small-engine electrical system is covered with an insulator. The insulation prevents the current from leaking out before it gets to its destination. Examples of materials which make good insulators are plastic and rubber.

Circuit. A circuit is a path or network of paths that will allow current to flow to do some work. Any circuit, no matter how complicated, is made up of several essential parts. A circuit is shown in Figure 8-4. There must always be a source of electrical pressure or voltage. In this illustration, the voltage source is a battery. In this circuit the current flow is used to light a light bulb. The light bulb will offer resistance to the current flow. A switch is necessary to turn the current flow on or off in the circuit. Wires or conductors connect the battery, switch and light bulb. Our circuit, then, has a voltage source (battery), a resistance unit (light bulb) and a switch connected by conductors (wires). In order for current to flow in a circuit, the path must be unbroken. In fact, the term *circuit* means *circle*.

Ground Circuit. Not only can electricity flow through wires, it also can flow through metal

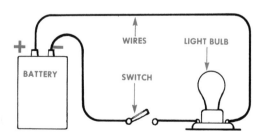

Figure 8-4. Parts of a circuit.

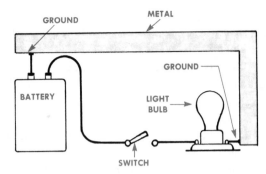

Figure 8-5. Electrical parts can be connected to metal instead of to a wire to make a complete circuit.

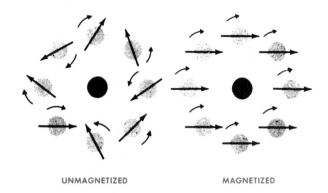

Figure 8-6. Magnetism is the alignment of electron orbits. (Clinton Engines Corp.)

parts on the engine. We can use these metal parts in place of one of the wires in a circuit. Look at the example in Figure 8-5. Instead of a wire from the light bulb to the battery, we could connect a short wire from the battery to a metal part on the engine. We could connect a short wire from the light to the same metal part. Electricity will flow across the closed switch into the light. Current will flow through the light into the metal on the engine and back to the battery. This completes the circuit and the light lights up. The part of the circuit connected to the metal is called *ground* or *grounded*. We call this a grounded circuit or ground circuit. Most electrical parts have one connection grounded. This means we only need one wire to have a complete circuit between two electrical parts.

MAGNETISM

Magnetism is a force that is involved in the operation of the magneto. The word *magneto,* in fact, comes from the word *magnet* or *magnetism.* Exactly what magnetism is and how it uses its force is still not completely understood. One theory is based upon the electron. This theory says that each electron has a circle of magnetic force around it. In an unmagnetized piece of iron, the electron orbits are not arranged in any pattern. In a magnetized piece of iron, the electron orbits are lined up so that their circles of magnetic force are added together, Figure 8-6. When all the magnetic forces work together, the piece of iron has a strong magnetic field.

INDUCTION

Induction involves the transfer of energy from one object to another without the objects touching each other. Induction is used in the magneto coil.

When current flows through a coil, a magnetic field is created in the coil. If the coil with current flowing in it is placed near another coil without current, the two coils will influence each other. If two coils are placed next to each other as shown in Figure 8-7, the bulb connected to the second coil will not light. If the switch is opened, the field around the first coil will collapse and jump over to the second coil. This collapsing magnetic field will cause current to flow in the second coil. The light connected to the second coil will light, but just for a fraction of a second. If the first coil is charged, then collapsed again, the light will light again. The kind of induction is used in magneto ignition coils.

MAGNETO MAGNETS

As we said earlier, the word *magneto* comes from the word *magnet.* All magnetos have magnets. Magnetism may be used to help make electricity. The magneto magnet is attached to the engine's flywheel. The magnet may be attached to the inside of the flywheel, as shown in Figure 8-8. On other engines it is attached to the outside of the flywheel. The magnet may be cast into the flywheel, or it may be mounted to the flywheel with bolts. The magnet goes around with the flywheel. We use this revolving magnet to make electricity.

Armature and Coil

We cannot make electricity with just a magnet. We need an armature and a coil. An armature and coil are shown in Figure 8-9. The armature is made from several thin strips of soft iron. The strips are squeezed tightly together. The armature is used to make a path for the magnetism.

The coil is attached to the armature. Inside the coil is a fairly thick wire. This wire is wrapped, or coiled, around part of the armature, as shown in Figure 8-10. It is called the *primary wire*. One end of the primary wire is attached to the armature. The other end goes to a switch called the *breaker points*. We will see how these work later.

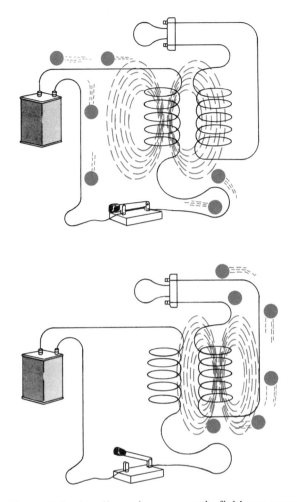

Figure 8-7. A collapse in a magnetic field can cause current to flow in another coil.

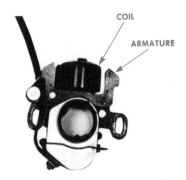

Figure 8-9. An armature and coil are part of the magneto.

Figure 8-8. A magnet is attached to the flywheel.

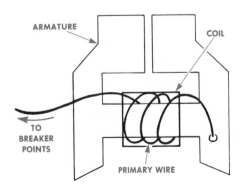

Figure 8-10. There is a thick wire, called a primary wire, inside the coil.

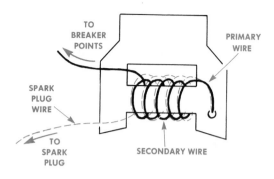

Figure 8-11. There is a thin wire in the coil, called the secondary wire.

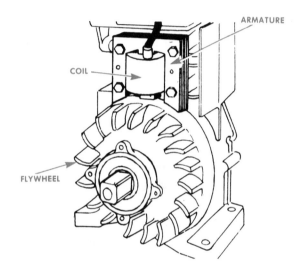

Figure 8-12. The armature and coil are mounted next to the flywheel and magnet. (Briggs & Stratton Corp.)

There is another wire inside the coil. It is called the *secondary wire*. The secondary wire is wrapped around the primary wire. The secondary wire is much thinner than the primary, and is wrapped around the armature many more times.

One end of the secondary wire is hooked to the armature. The other end is hooked to the thick wire that goes to the spark plug, Figure 8-11.

The armature and coil are mounted next to the flywheel. In Figure 8-12 the magnet is attached to the outside of the flywheel. The armature and coil are mounted right above the flywheel. On some

engines, the magnet is mounted inside the flywheel, and the coil and armature are mounted under the flywheel.

Magnet and Coil Operation

Let's see how a magneto makes electricity. The magnet goes around with the flywheel. As the flywheel turns, the magnet goes under the armature, as shown in Figure 8-13. Magnetism from the magnet moves from one end of the magnet and into the armature. The magnetism travels through the armature and back into the magnet.

The flywheel turns some more. The magnet lines up under another part of the armature, as shown in Figure 8-14. The magnetism now passes through the armature in the opposite direction. When magnetism changes direction next to a coil, a small amount of electricity builds up in the coil's primary wire.

The amount of electricity we get in the primary wire is not enough to get the air-fuel mixture in the cylinder burning. To burn the air-fuel mixture we need a lot of electricity. We have to change the small amount of electricity in the primary wire to

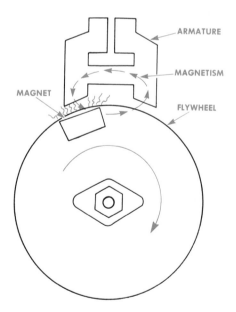

Figure 8-13. As the magnet lines up under the armature, magnetism goes in one direction.

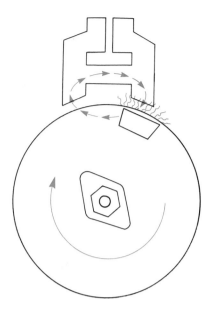

Figure 8-14. As the magnet turns, more magnetism goes in the other direction.

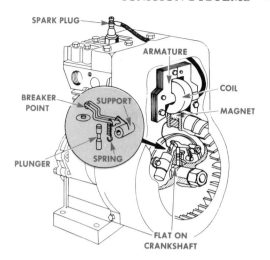

Figure 8-15. The breaker points fit behind the flywheel. (Briggs & Stratton Corp.)

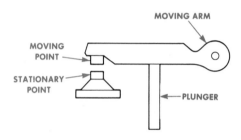

Figure 8-16. The parts of the breaker points.

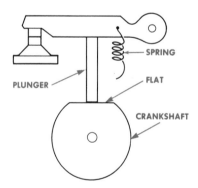

Figure 8-17. A flat spot on the crankshaft allows the breaker points to close.

high-voltage electricity in the secondary wire. To do this we need some more parts. We will continue our study of the magneto operation in the next section.

BREAKER POINTS

The breaker points are a switch that is opened and closed by the crankshaft or camshaft. The breaker points are usually under the flywheel, as shown in Figure 8-15.

The breaker points are two small, round points. One of the points is called the *moving point*. It is connected to a moving arm. A small, round rod is used to move the arm. This rod is called a *plunger*. The other point does not move. It is called the *stationary point*, Figure 8-16.

The breaker points fit right next to the crankshaft. The crankshaft opens and closes the breaker points. The plunger rides on the crankshaft. The crankshaft has a small, flat spot on it. When the plunger is on the flat spot, the contact points are closed. A small spring helps hold the points closed. Closed breaker points are shown in Figure 8-17.

As the crankshaft turns, the flat spot moves away from the plunger. The plunger pushes up on the moving point arm. The breaker points open, as shown in Figure 8-18.

Some engines have breaker points that are

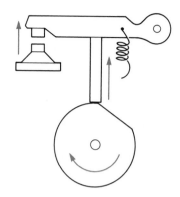

Figure 8-18. The round part of the crankshaft pushes the plunger up and opens the breaker points.

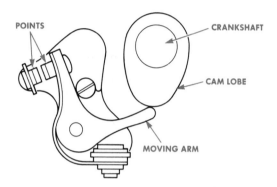

Figure 8-19. A cam can be used to open the breaker points.

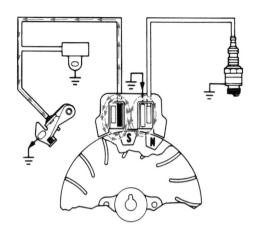

Figure 8-20. When the breaker points are closed, electricity flows in the primary wire. (McCulloch Corp.)

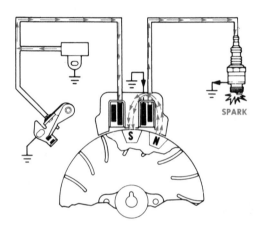

Figure 8-21. When the breaker points open, high-voltage electricity flows in the secondary wire. (McCulloch Corp.)

opened by a cam. A cam lobe is attached to the crankshaft. The cam turns with the crankshaft. When the cam lobe hits the moving arm, the breaker points are opened, Figure 8-19.

Breaker Point Operation

Now let's see how the breaker points work to help build high-voltage electricity. The magnet on the flywheel generates a small amount of electricity as it goes by the armature. The electricity is made in the primary wire.

One end of the primary wire is attached to the armature. The other end is attached to the moving point arm. Electricity can flow through the primary wire and into the moving arm. The breaker points are closed. Electricity can now flow across

from the moving point to the stationary point, Figure 8-20.

When the flywheel has turned a little more, the breaker points open, Figure 8-21. Electricity cannot flow through the open points. The electricity flowing in the primary wire stops quickly. This causes a flow of magnetism to rush through the secondary wire. The magnetic flow creates a very high-voltage electricity in the secondary wire. This voltage may be as high as 25,000 volts. This is enough electricity to get our air-fuel mixture burning.

THE CONDENSER

Electricity must stop flowing the instant the breaker points open. If it does not, we cannot build up high-voltage electricity. Electricity will try to arc, or jump across the open breaker points, as shown in Figure 8-22. If this happens, we do not get a magnetic flow in the coil. We cannot build up a high-voltage electricity.

Electricity jumping across the breaker points causes another problem. The points will be burned and soon be destroyed. They would no longer work.

A condenser is used to stop electricity from jumping across open points. A condenser is a small electrical part shaped like a tiny can. It has a small wire that comes out of one end. A little bracket allows it to be attached to the magneto.

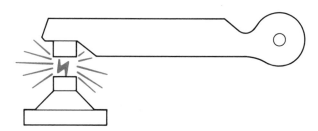

Figure 8-22. When the breaker points open, electricity tries to jump across the points.

The condenser is mounted next to the breaker points. The condenser wire is hooked to the moving point. A condenser and breaker points are shown in Figure 8-23.

A condenser is shown in cross section in Figure 8-24. Inside, the condenser consists of two long sheets of conductor foil separated by several sheets of insulated paper. The foil and the insulation are wound into a tight roll. The roll of insulated and conductor paper is installed in a small metal canister. An insulated end-piece through which a small insulated wire lead, or pigtail, is routed, is placed on top of the canister. The end of the canister is crimped over the insulated end-piece. A gasket is located between the end-piece and the foil sheets. A spring sometimes is used at the bottom of the canister to maintain pressure on the end-piece gasket.

Very small amounts of moisture can have a very bad effect on the paper insulation inside a condenser. Moisture can lead to early condenser failure. For this reason, air and moisture are removed with heat and vacuum in a process known as hermetric sealing. The spring, gasket and tight-fitting end-piece are designed to maintain the hermetic seal through the service life of the condenser.

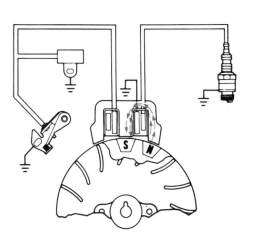

Figure 8-23. The condenser wire is hooked to the moving point. (McCulloch Corp.)

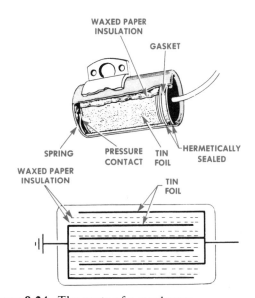

Figure 8-24. The parts of a condenser.

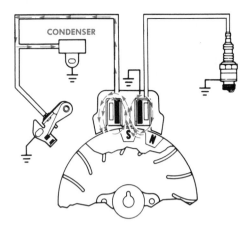

Figure 8-25. When the points open, electricity flows into the condenser. (McCulloch Corp.)

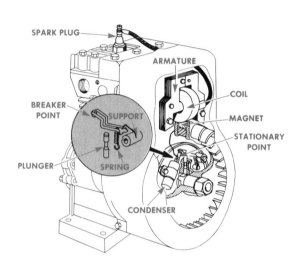

Figure 8-26. The condenser and stationary point may be one part. (Briggs & Stratton Corp.)

The condenser works like a tiny storage tank. The breaker points open. Electricity wants to jump across the opening. The electricity has another place to go. It goes through the condenser wire and into the condenser, Figure 8-25.

The condenser stores electricity inside. Electricity does not jump across the open points. The points do not burn as quickly.

Some engines have a condenser without a wire. The stationary point is hooked directly to the condenser. An engine with this type of one-piece condenser is shown in Figure 8-26.

IGNITION CABLES

The high voltage developed in the coil's secondary wire must be sent to the spark plug. The wire used for this job is called a secondary wire high-voltage wire, spark plug wire, high tension wire or ignition cable. Ignition cables must be able to handle high voltage without leakage. They must also be able to withstand water, oil, vibration and abrasion. Usually the wire is soldered into the coil at one end. It has a terminal connection for the spark plug at the other end, Figure 8-27.

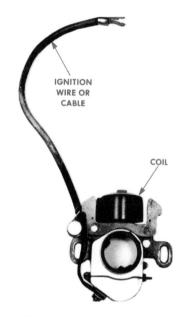

Figure 8-27. The high voltage from the coil goes to the spark plug through an ignition cable.

SPARK PLUG

The spark plug gets the high-voltage electricity from the magneto. Its job is to make a spark to get the air-fuel mixture burning. The outside parts of a spark plug are shown in Figure 8-28. The top of the spark plug has a terminal. This is where the spark plug wire is connected. The terminal is connected to a wire that goes through the middle of the spark plug. This is the wire that allows high-

Figure 8-28. The parts of a spark plug. (Champion Spark Plug Co.)

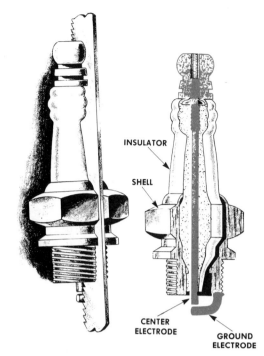

Figure 8-29. Sectional view of a spark plug. (General Motors Corp.)

voltage electricity to get into the combustion chamber.

The high-voltage electricity must not be allowed to leak away. An insulator fits around the wire. The insulator makes sure the electricity goes through the middle of the spark plug. A metal part called a shell makes up the bottom of the spark plug. The shell has threads. The threads allow the spark plug to be screwed into the cylinder head. The shell also gives us a place to fit a wrench. Hex shaped flats on the outside of the shell allow a wrench to be used on the spark plug for installation and removal.

Basically, a spark plug is a wire with an air gap at the bottom, that will fit into the engine's combustion chamber. A sectional view of a spark plug is shown in Figure 8-29. The wire which conducts high voltage into the cylinder is called the center electrode. There is terminal at the top of the center electrode to attach a connector from an ignition cable.

Since the center electrode must carry high voltage into the cylinder, it must be well insulated. A

ceramic insulator surrounds the center electrode. The ceramic insulator has ribs on its outside diameter to increase the distance between the terminal and the nearest ground. This helps eliminate current leakage, or flashover, especially when the outside of the ceramic is dirty or wet.

The center electrode and ceramic insulator assembly are joined to a metal shell. The shell, insulated from the center electrode by the ceramic, has threads rolled on it to allow the spark plug to be screwed into the combustion chamber. A side electrode is attached to the shell and placed a small distance away from the center electrode. This distance is the air gap or spark plug gap that the current jumps to create a spark.

The spark plug, mounted in the combustion chamber, is subjected to extremely high pressure. Seals are used between the shell and ceramic insulator and between the center electrode and the ceramic to prevent the leakage of combustion pressure. Either a copper gasket or a special taper seat is used to prevent leakage of combustion pressure around the shell threads.

Spark Plug Operation

Secondary high voltage flows from the magneto coil through the high-voltage ignition cables. The voltage enters the spark plug at the terminal end of the center electrode. Voltage flows down the center electrode to the air gap located in the engine's combustion chamber. The current overcomes the resistance of the air-fuel mixture and jumps the air gap to the side electrode. The spark created as the current jumps the gap ignites the combustible mixture of air and fuel.

The voltage required to overcome the gap and the resistance of the air-fuel mixture is different under different conditions. The wider the air gap, the higher the required voltage. The condition of the spark plug electrodes also is very important. Much less voltage is required to jump from clean, sharp electrodes than from dirty, eroded ones. The higher the compression pressure, the higher the voltage required to overcome the air gap.

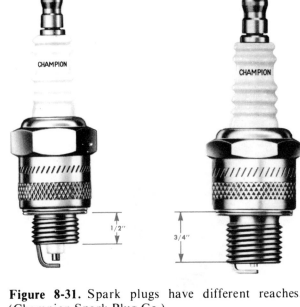

Figure 8-31. Spark plugs have different reaches. (Champion Spark Plug Co.)

Spark Plug Sizes

If it is to work properly, the spark plug must be the right size for the combustion chamber into which it is installed. Different sizes of spark plugs are required for different engine designs. Spark plugs are made with different shell thread diameters, Figure 8-30. The threads are made in metric sizes measured in millimeters. The metric thread allows the plugs to be used in both imported and American engines.

The threaded section of the shell is made in different lengths. This dimension is called the *reach*. There are several common reach dimensions manufactured. Figure 8-31 shows two. The

Figure 8-30. Spark plugs are made with different thread sizes. (Champion Spark Plug Co.)

thickness of the combustion chamber determines what reach is necessary.

Heat Range

The spark plug tip, or electrode area, mounted in the engine's combustion chamber, is subjected to temperatures that may exceed 2,000° F. The firing end of the spark plug is designed to remove this heat through the engine's cylinder head. The path of heat flow away from the firing end is shown in Figure 8-32. Heat moves up the ceramic insulator to the metal shell and then out into the engine's cylinder head.

Spark plugs are designed to operate within a specific temperature range. The term *heat range* describes the ability of a spark plug to conduct heat away from the firing end. The heat range of a spark plug is determined by the distance the heat must flow from the firing end to the shell. This is determined by the length of the insulator firing end. If the path is a long one, the firing end will remain at a high temperature and it is referred to as a *hot spark plug*. If the path for heat flow is short, heat is removed more easily from the firing end and the spark plug is cooler in operation. This type of spark plug is referred to as a *cold spark plug*. The heat paths for a cold and a hot spark plug are shown in Figure 8-33.

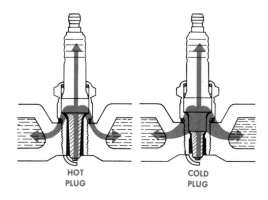

Figure 8-33. Heat flow in a hot and cold spark plug. (NGK)

Electrode Design

Several different kinds of electrode designs are used with small engines. The common electrode designs are shown in Figure 8-34. The automotive-type electrodes are used in many four-stroke lawn mower and motorcycle engines. The dual gap design is used in many two-stroke-cycle engines where fouling is a potential problem. The retracted gap also is popular in two-stroke engines. The surface gap is used with many solid-state magneto systems.

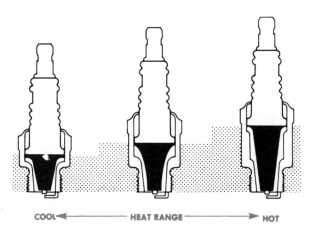

Figure 8-32. Heat flow away from a spark plug firing end.

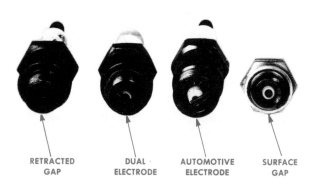

Figure 8-34. Types of spark plug electrodes.

Spark Plug Codes

In order to be correctly matched to an engine, a spark plug must have the correct thread diameter, gasket or tapered seat, heat range, electrode type and reach. Spark plug manufacturers identify these items on their spark plugs with a code system. The code is printed on the ceramic insulator of the spark plug, Figure 8-35.

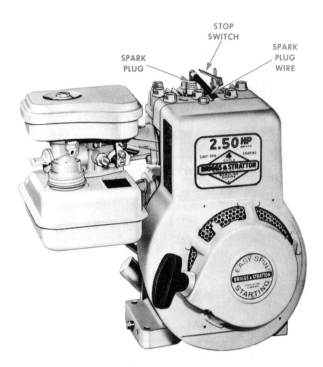

Figure 8-36. When the stop switch is pushed against the spark plug terminal, the secondary is grounded. (Briggs & Stratton Corp.)

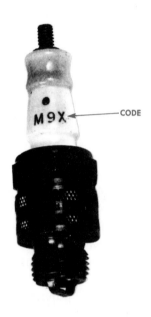

Figure 8-35. Spark plugs are matched to the engine by a code on the insulator.

STOP SWITCH

There must be a way to stop a running engine. Engines are stopped by a stop switch. The stop switch opens the circuit and grounds out the ignition system. The grounding may be done at the magneto primary wire or at the spark plug wire. On many engines, a wire is connected to the movable contact point. This wire runs outside the engine to a grounded switch. When the switch is closed, primary magneto current cannot flow and the engine stops.

Another popular stop switch consists of a thin metal strip attached to the top of the engine,

Figure 8-36. When the engine is to be stopped, the metal is pushed against the spark plug terminal. This grounds out the ignition secondary current and stops the engine.

CAPACITIVE DISCHARGE IGNITION

An ignition system that uses contact points has some limitations. Contact points wear as the engine runs. Eventually the contact points will be so worn that the engine will be hard to start and will not run well. The contact points must be changed to get the engine back into running condition.

A new type of ignition system is now being used on some small engines. This system, called the *capacitive discharge* or *CD system*, does not use contact points in the magneto. The capacitive discharge ignition system uses the principle of storing and discharging energy from a condenser or capacitor.

As was described earlier under conventional ignition systems, a condenser (or capacitor) is made of two parallel plates separated by an insulator. When current enters the capacitor, electrons build up on one plate, and their negative charge repels a like number of electrons on the other plate. In this condition, the capacitor is said to be charged. Energy is stored in the capacitor; when the current flow is stopped, the energy remains in the capacitor. Only when a conductor is connected across the two plates will it discharge, or regain electron balance. A small capacitor is capable of storing a large electron charge and providing a big discharge.

In a capacitive discharge system, a charged capacitor is placed across the primary winding of an ignition coil. As the capacitor discharges into the primary winding, a strong magnetic field is established and cuts the secondary winding, inducing a high voltage. Energy is not stored in the coil; it is used only to step up the voltage from the capacitor. The energy developed in this way is greater than that possible in a conventional ignition system. The capacitor is then disconnected from the coil and recharged, so that the discharge into the primary can occur again for the next firing cycle.

The parts of a CD magneto ignition system are shown in Figure 8-37. A simplified diagram of the system is shown in Figure 8-38. Magnets on the

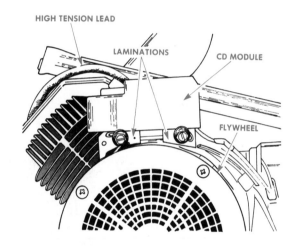

Figure 8-37. Parts of a CD magneto system.

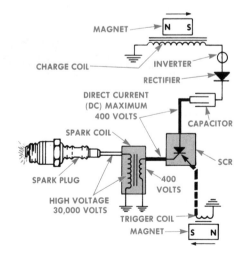

Figure 8-38. A CD ignition system.

flywheel are used just as in a conventional system. As they move under the armature, a small current is developed in a coil called the *charge coil*.

It is necessary to use a device called an *inverter* to quickly build up a charge in the energy storage capacitor. The inverter changes the low-voltage direct current available from the charge coil into alternating current. The alternating current transformed by the inverter is changed back to direct current by the bridge rectifier. The output of the inverter and rectifier needed to charge the energy storage capacitor is approximately 400 volts.

The 400 volts stored in the capacitor must be discharged into the coil primary at just the right time for ignition. The flywheel magnets are used to develop a small signal current in a coil called the *trigger coil*. The signal from the trigger coil goes to a switching device called a *silicon controlled rectifier* (SCR). The SCR gets the signal from the trigger coil and switches the circuit to cause the capacitor to discharge into the spark coil primary. A high voltage induced in the spark coil is directed to the spark plug. The SCR then switches the circuitry back, to allow the capacitor to charge, to get ready for the next discharge.

NEW TERMS

atoms: Small particles which make up matter.

breaker points: The switch used in the ignition primary system to control coil operation.

capacitive discharge ignition system: An ignition system that uses the energy stored in a capacitor to develop high voltage.

capacitor: An electrical device used to store or soak up a surge of electricity.

circuit: A complete path for electrical current flow.

coil: An electrical device used to step up voltage for ignition.

condenser: The capacitor used in the ignition primary to prevent contact breaker point arcing.

conductor: A material that allows electrical current flow.

current: The flow of electrons in an electrical circuit. Measured in amperes. Abbreviated *I*.

electricity: The flow of electrons from one atom to another.

ignition cables: High-voltage ignition wires used to carry secondary voltage.

ignition system: The electrical system that provides the high-voltage spark to ignite the air-fuel mixture in the cylinder.

induction: The transfer of energy from one object to another without the objects touching.

insulator: A material that prevents the flow of electricity.

magneto: Device used to develop the high voltage necessary for ignition.

spark plug: Ignition-system part used to create a spark in the combustion chamber.

voltage: The source of potential energy in an electrical system. Measured in volts and abbreviated *E*.

SELF CHECK

1. Write a definition for *electricity*.
2. Where are the breaker points on an engine?
3. Where does electricity flow when the breaker points are closed?
4. What happens to the flow of electricity when the breaker points open?
5. What happens in the secondary wire when the breaker points open?
6. Where does the spark plug fit?
7. What wire is connected to the spark plug?
8. How does electricity go through the spark plug?
9. How does a spark plug make a spark?
10. How is a CD ignition system different from one with breaker points?

DISCUSSION TOPICS AND ACTIVITIES

1. Use a small engine cutaway model to identify the parts of a magneto ignition system.
2. Turn an engine's crankshaft and watch the breaker points open and close. Can you describe the magneto operation?

unit 9
fuel systems

The engine operates on a mixture of air and fuel. The fuel system has two main parts, as shown in Figure 9-1. The fuel tank or gas tank allows the engine to store enough fuel for several hours of operation. The engine part used to mix air and fuel together is called a *carburetor*. There are many types of carburetors. They all work in the same basic way. In this unit we will study the operation of the fuel system.

LET'S FIND OUT: When you finish reading and studying this unit, you should be able to:
1. Explain the parts and operation of a basic carburetor.
2. Describe the parts and operation of a vacuum or suction carburetor.
3. Describe the parts and operation of a float carburetor.
4. Describe the parts and operation of a diaphragm carburetor.
5. Explain the operation of a sliding valve carburetor.

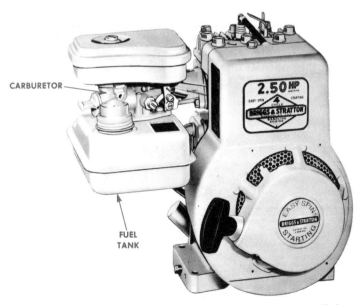

Figure 9-1. The two main parts of the fuel system are fuel tank and carburetor. (Briggs & Stratton Corp.)

THE CARBURETOR

One of the engine's strokes is called the *intake stroke*. On this stroke the piston moves down very fast. At the same time, the intake valve is opened. This fast downward movement causes a vacuum in the cylinder, and this is how the air-fuel mixture gets into the cylinder.

The carburetor's job is to mix the air and fuel together in just the right amount. Next to the intake valve, the carburetor is attached to the engine. A carburetor is shaped like a tube, as shown in Figure 9-2. The tube is open at one end. When the piston is on the intake stroke, air is pulled in the open end of the carburetor. The air can get into the engine through the open intake valve.

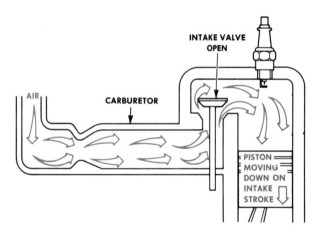

Figure 9-2. A carburetor is a tube open at one end, attached to the engine.

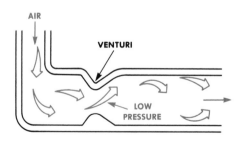

Figure 9-3. The venturi is a raised part of the carburetor that causes a low-pressure area.

Venturi

The middle of our carburetor tube has a partially restricted area. This area is called the *venturi*, Figure 9-3. Air has a hard time getting through the venturi. As the air does get through, it speeds up, leaving low pressure in the venturi. This low pressure will pull in fuel.

Fuel Pickup

We must have a way of mixing fuel with the air. A small amount of fuel is stored in the carburetor, as shown in Figure 9-4. The fuel is right under the venturi. A hollow pickup tube runs from the fuel to the venturi.

When air goes through the venturi, it causes a low pressure. This low pressure sucks fuel up the pickup tube. The fuel mixes with the air. The mixture of air and fuel goes into the engine.

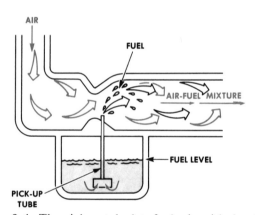

Figure 9-4. The pickup tube lets fuel mix with the air.

Throttle Valve

There are times when we want an engine to run fast. At other times we want it to run slowly. If a large amount of air-fuel mixture enters, the engine runs fast. If just a little air-fuel mixture gets in, the engine will run slowly.

We use a part called the *throttle valve* to control engine speed. A throttle valve is shown in Figure 9-5. The throttle valve is a round plate. It fits in the end of the carburetor tube. The valve may be opened or closed by the person using the engine.

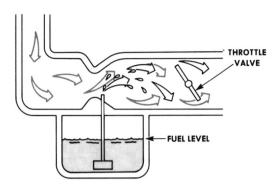

Figure 9-5. The throttle valve fits in the end of the carburetor.

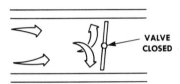

Figure 9-6. A closed throttle valve prevents air and fuel from getting into the engine.

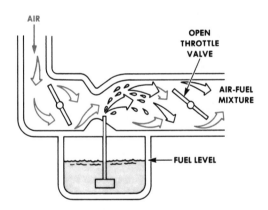

Figure 9-7. An open throttle valve lets in a large amount of air-fuel mixture.

A closed valve is shown in Figure 9-6. Very little air and fuel can get into the engine, so the engine runs slowly. The more the valve is opened, the more air-fuel mixture gets into the engine, and the faster the engine runs. An open throttle valve is shown in Figure 9-7.

Choke

When an engine is cold, it is hard to start. A carburetor has a choke, Figure 9-8, which helps start a cold engine.

The choke is a small round valve. It looks a lot like the throttle valve. The choke valve fits in the carburetor near where the air comes in.

When the engine is cold, we close the choke valve. Very little air can get into the carburetor. The piston going down causes a vacuum. This strong vacuum pulls a lot of fuel out of the pickup tube. The fuel goes into the cylinder. All this fuel helps the engine start more easily.

The choke valve usually is connected to a lever, Figure 9-9. The lever is on the outside of the carburetor. We use the level to close the choke valve when the engine is cold. We move the lever to open the choke valve when the engine is warm.

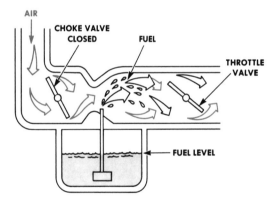

Figure 9-8. When the choke valve is closed, a lot of fuel is pulled up the pickup tube.

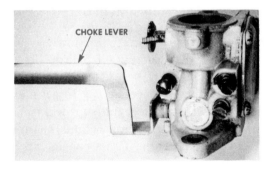

Figure 9-9. A choke lever is used to open and close the choke valve.

High-Speed Fuel Adjustment Screw

When the engine runs at high speed, it needs a lot of air and fuel. Air enters the carburetor and goes through the venturi. A vacuum in the venturi pulls fuel up the pickup tube. The fuel mixes with the air and enters the cylinder.

We need a way of controlling the amount of fuel that goes up the pickup tube. A screw with a pointed end fits in the side of the carburetor as shown in Figure 9-10. The end of the screw goes into the pickup tube. We use the screw to open or close the pickup tube. If we turn the screw all the way in, not much fuel can go up the tube. If we turn the screw out, a lot of fuel can go up the tube.

This screw is called the high-speed fuel adjustment screw. We use it to adjust the amount of fuel going into the engine at high speed. In a later unit we will see how to adjust this screw.

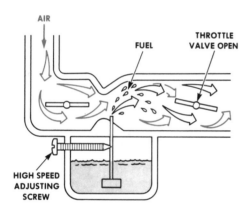

Figure 9-10. The high-speed adjustment screw controls the fuel coming up the pickup tube.

Low-Speed Fuel Adjustment Screw

When we want an engine to run slowly, we close the throttle valve, so that very little air goes through the venturi and very little fuel is pulled up the pickup tube. We need a way to get a small amount of fuel into the engine so it will run. We do this by having a small hole behind the throttle valve. Fuel is pulled through a passage and out the hole as shown in Figure 9-11.

We need a way to control the amount of fuel that comes out the hole. A small screw with a

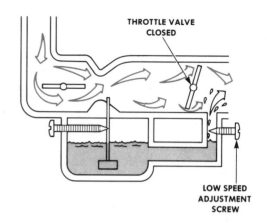

Figure 9-11. The low-speed screw controls the fuel coming out behind the throttle valve.

Figure 9-12. Low-speed adjustment screw.

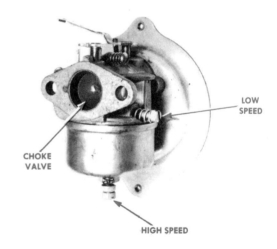

Figure 9-13. The low-speed and high-speed adjustment screws and the choke valve can be seen in this carburetor.

pointed end is used, Figure 9-12. The screw goes through the side of the carburetor. The pointed end goes into the fuel passage. If the screw is turned all the way in, very little fuel can come out the hole. If the screw is turned out, more fuel can get out the hole. We will learn how to adjust this screw in a later unit. The carburetor shown in Figure 9-13 has the low-speed fuel adjustment screw, the high-speed fuel adjustment screw and the choke valve identified.

SUCTION CARBURETOR

The suction or vacuum carburetor is one of the most common carburetors. Many small engines use a vacuum carburetor. A vacuum carburetor is mounted on top of a fuel tank. Fuel is pulled out of the fuel tank by a vacuum. All vacuum carburetors work the same. They are sometimes called vacu-feed or vacu-jet carburetors.

Suction Carburetor Parts

A vacuum carburetor is shown in Figure 9-14. The carburetor has two bolt holes used to mount it to the engine. There is an opening at the top of the carburetor, which allows air to enter. The air goes through the center of the carburetor. A choke is mounted next to the air opening. The choke can be used to open or close the air opening.

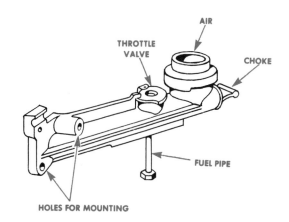

Figure 9-15. Parts of a vacuum carburetor. (Briggs & Stratton Corp.)

A throttle valve is inside the carburetor. Right below the throttle valve is a tube called the *fuel pipe*, which brings fuel up into the carburetor. A fuel pipe is shown in Figure 9-15. There is a small screen in the bottom of the pipe. The screen stops dirt from going up the pipe and into the carburetor. A small ball fits in the bottom of the pipe, Figure 9-16. The ball lets fuel go up the pipe, but it will not let fuel run back out of the pipe. If the pipe is empty, it takes a long time to start the engine.

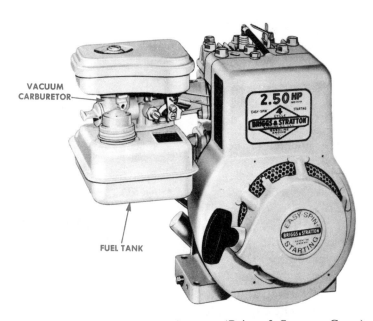

Figure 9-14. An engine with a suction or vacuum carburetor. (Briggs & Stratton Corp.)

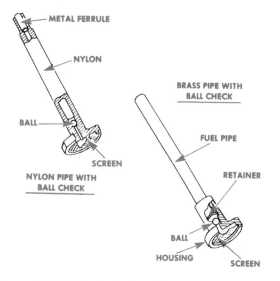

Figure 9-16. The fuel pipe has a screen and ball. (Briggs & Stratton Corp.)

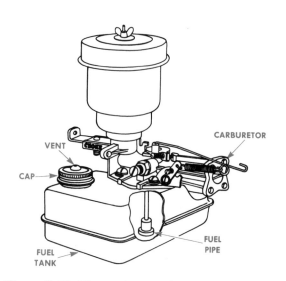

Figure 9-17. The carburetor is mounted to the fuel tank. (Briggs & Stratton Corp.)

The fuel tank is attached to the bottom of the carburetor. A fuel tank and carburetor are shown in Figure 9-17. The fuel tank is used to store fuel. There is enough gasoline in the tank to allow the engine to run for several hours. The cap on the fuel tank has a small hole in it, called a *vent*. The

vent allows air to get into the tank. Without a vent, a vacuum could form in the tank and prevent fuel from going up the fuel pipe. The fuel pipe on the carburetor fits down into the fuel tank.

How the Suction Carburetor Works

As the piston moves down in the cylinder, a vacuum is created inside the carburetor. A vacuum inside the carburetor pulls fuel up the fuel pipe. The amount of fuel is controlled by a high-speed adjustment screw, shown in Figure 9-18. The throttle valve slows down the air much like a venturi; this helps pull fuel up the pipe. There are

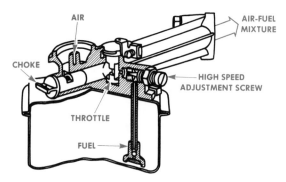

Figure 9-18. Air going through the carburetor causes a vacuum that pulls fuel out of the fuel tank. (Briggs & Stratton Corp.)

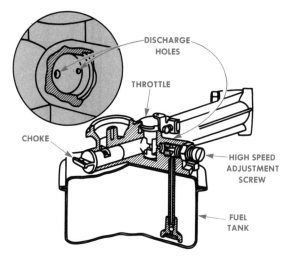

Figure 9-19. Fuel enters through two discharge or metering holes. (Briggs & Stratton Corp.)

two holes through which the fuel enters the carburetor. These are called *metering* or *discharge holes,* Figure 9-19. When the throttle is open, fuel comes out both holes. When the throttle is closed, a small amount of fuel can come out one of the

holes. This allows the engine to run with a closed throttle.

The choke is a sliding plate mounted at the outer end of the carburetor, Figure 9-20. The choke is pulled out to close off the air. This helps in starting a cold engine. The choke is pushed in as soon as the engine starts.

FLOAT CARBURETOR

Many small engines use a float carburetor, Figure 9-21. A separate fuel tank is used with a float carburetor. The fuel tank is attached to another part of the engine. The fuel tank is mounted higher than the carburetor, as shown in Figure 9-22. Fuel flows from the tank through a fuel line to the carburetor. Gravity makes the fuel flow. A float in the carburetor controls the flow of fuel from the tank.

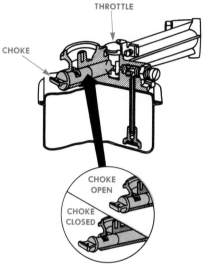

Figure 9-20. The choke is opened to let in air and closed to stop the air. (Briggs & Stratton Corp.)

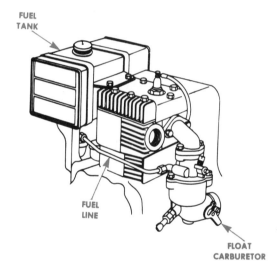

Figure 9-22. The fuel tank is mounted above a float carburetor. (Briggs & Stratton Corp.)

Figure 9-21. An engine with a float carburetor. (Briggs & Stratton Corp.)

The Float

The fuel line from the fuel tank brings fuel into the carburetor. The fuel line is connected to a hole, called the *fuel inlet,* in the side of the carburetor. The fuel goes into an area called the *float bowl.* An inlet and float bowl are shown in Figure

Figure 9-23. A float carburetor.

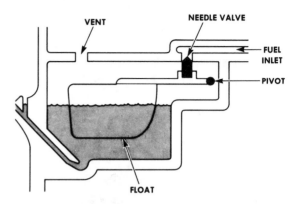

Figure 9-25. Parts of the float system.

Figure 9-24. A float and needle valve.

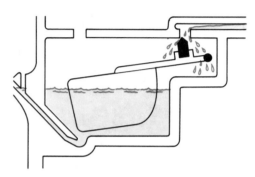

Figure 9-26. When the float is down, the needle valve lets fuel come in.

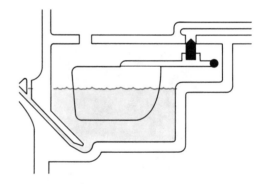

Figure 9-27. As the fuel level goes up, the float goes up and shuts off the fuel.

9-23. Fuel in the float bowl is used to run the engine.

The amount of fuel in the float bowl is controlled by a float. The float is shown in Figure 9-24. It is a small hollow piece of copper or plastic. The float is light enough to float on top of gasoline. The float is connected to a pivot on the side of the bowl. A small, round piece of metal with a sharp end fits on top of the float. This part is called a *needle valve*, Figure 9-25. The needle valve fits into the fuel inlet. There is a small vent in the top of the bowl to let in air and equalize pressure in the float chamber.

As the engine runs, some of the fuel in the float bowl is used up. As the fuel level in the bowl drops, the float drops. As the float moves down, the needle valve is moved out of the inlet hole. Fuel can come in through the inlet hole from the fuel tank, as shown in Figure 9-26.

The fuel level will rise as more fuel comes into the bowl. The float will also rise, pushing the needle valve into the inlet, as shown in Figure 9-27. Now fuel cannot get into the bowl around the needle valve. Flow from the fuel tank is stopped.

Float Carburetor Parts

All float carburetors use a float to control the fuel level. A float carburetor has the same basic parts as other carburetors we have studied. Two types of float carburetors are shown in Figure 9-28. Many float carburetors have a low- and a high-speed adjustment screw. They have a throttle valve and choke valve.

The inside parts of a float carburetor are shown in Figure 9-29. There is a tube with holes in it. It brings fuel from the float bowl to the venturi. The tube is called a *nozzle*. It does the same job as the fuel pipe in a suction carburetor.

How The Float Carburetor Works

When the engine is running fast, the throttle valve is open as shown in Figure 9-30. Air comes

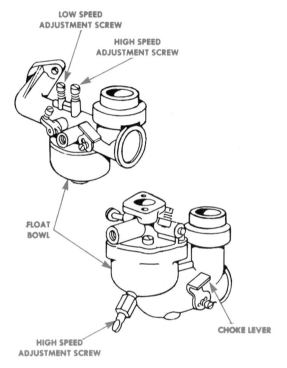

Figure 9-28. Parts on the outside of two float carburetors. (Briggs & Stratton Corp.)

Figure 9-29. Parts inside a float carburetor.

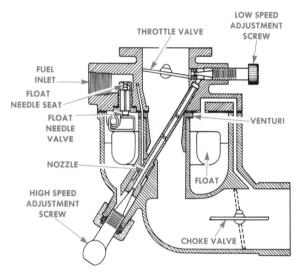

Figure 9-30. Air flowing through the venturi causes fuel to be pulled through the nozzle.

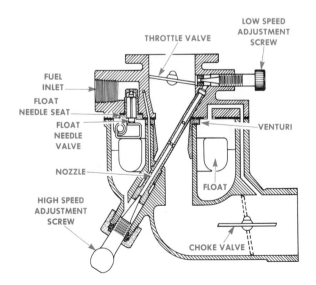

in the carburetor. As air goes through the venturi, low pressure is developed. Fuel is pulled out of the float bowl through the nozzle. Air and fuel mix together in the venturi. The air-fuel mixture enters the engine cylinder. The amount of fuel that can go up the fuel nozzle is controlled by the high-speed adjustment screw. Air bleed holes in the nozzle help break up the fuel on its way out of the nozzle.

When the engine is running slowly, the throttle valve is closed. There is a small hole on the engine side of the throttle valve. A vacuum at this hole causes fuel to be pulled up the nozzle. A small amount of fuel goes out of this hole and into the engine as shown in Figure 9-31. This is enough fuel for the engine to run.

A choke valve is used for cold starting. The choke valve is closed as shown in Figure 9-32. Not much air can enter the carburetor. A strong vacuum is caused by the engine's piston. This strong vacuum pulls a lot of fuel out of the nozzle. The fuel goes into the engine to help it start.

DIAPHRAGM CARBURETOR

Another kind of carburetor is used on many small engines. It is called a *diaphragm carburetor.* Float and suction carburetors work only on engines that are used upright. If a float or suction engine is turned on its side, the carburetor will not work properly. Some engines are used in many positions. For example, a chainsaw, Figure 9-33, must work in any position. A diaphragm carburetor is made to work in any position.

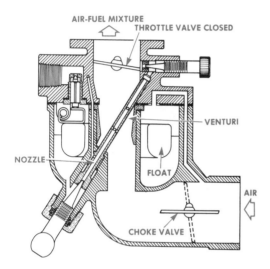

Figure 9-31. Fuel comes out of a hole behind the closed throttle valve.

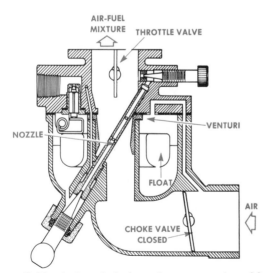

Figure 9-32. A closed choke valve causes a lot of fuel to be pulled out of the nozzle.

Figure 9-33. A chainsaw needs a diaphragm carburetor so it can work in all positions. (McCulloch Corp.)

Diaphragm Carburetor Parts

The outside parts of a diaphragm carburetor are shown in Figure 9-34. The top part of the carburetor is the same as a float or suction carburetor. It has a choke valve, throttle valve and venturi. The diaphragm carburetor has a high- and low-speed adjustment screw, Figure 9-35, just like other carburetors.

This kind of carburetor does not have a float bowl. Instead, it has a part called a *diaphragm*. The diaphragm controls a small amount of fuel in a fuel chamber. A fuel inlet allows fuel to enter the carburetor.

How a Diaphragm Carburetor Works

The diaphragm is made from a flexible rubber-like material. The diaphragm is stretched across a small space called a *chamber*, Figure 9-36.

A small control lever is connected from the center of the diaphragm to a needle valve. The needle valve has the same job as the needle valve in a float carburetor. A small spring fits between the top of the chamber and the control lever. The control lever is connected to a small pivot.

The fuel flows from the fuel tank to the fuel inlet. The spring pushes down on the control lever. This allows the needle valve to drop down.

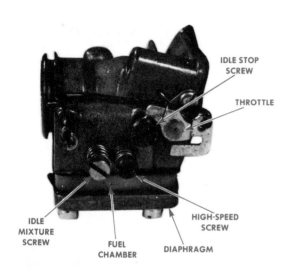

Figure 9-34. The outside parts of a diaphragm carburetor.

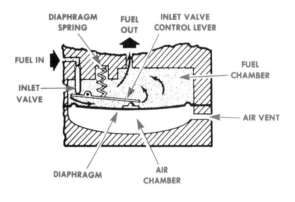

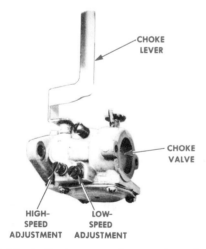

Figure 9-35. A diaphragm carburetor has a low- and a high-speed adjustment screw like other carburetors.

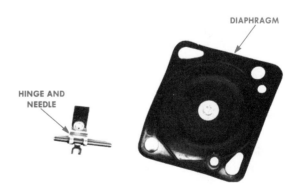

Figure 9-36. Diaphragm parts. (McCulloch Corp.)

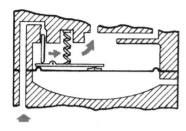

Figure 9-37. Fuel comes into the chamber around the needle valve.

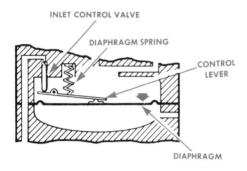

Figure 9-38. When the chamber is full, the diaphragm moves down and pushes the needle valve up.

Fuel can come in around the needle valve, as shown in Figure 9-37.

As the fuel fills up the chamber, it pushes down on the diaphragm. Downward movement of the diaphragm causes the control lever to pivot upward. This pushes up on the needle valve. The fuel inlet is now closed, as shown in Figure 9-38. When fuel is used up, the diaphragm comes back up. The needle will open to let fuel in again.

The diaphragm will work the same way no matter what position the engine is in. The engine can run even if it is upside down.

The rest of the carburetor works in the same manner as a float or suction carburetor. Air enters the carburetor and flows over a venturi. Low pressure in the venturi allows fuel to be pulled out of the fuel chamber. Fuel mixes with the air and enters the engine cylinder. An exploded view of a diaphragm carburetor is shown in Figure 9-39.

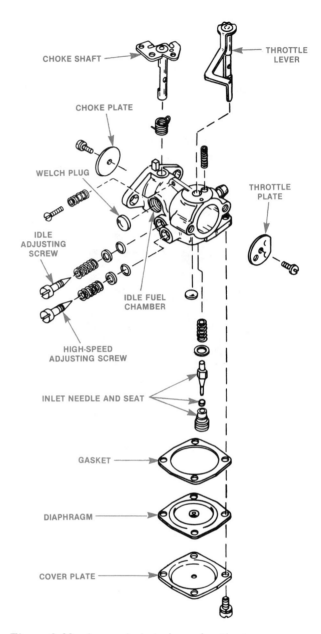

Figure 9-39. An exploded view of a diaphragm carburetor. (Tecumseh Products Co.)

SLIDING VALVE CARBURETOR

Many moped and motorcycle engines use a carburetor called a *sliding valve*. The carburetor gets its name from a piston shaped valve that slides up and down inside the carburetor, Figure 9-40. On the end of the sliding valve is a long needle valve. The valve controls the amount of

Figure 9-40. A sliding valve carburetor has a piston-shaped valve that slides up and down.

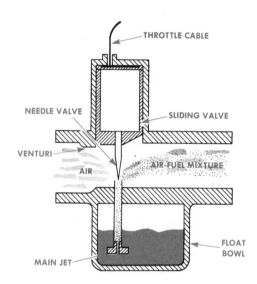

Figure 9-41. A cross-sectional view of a sliding valve carburetor.

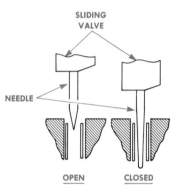

Figure 9-42. The needle is pulled out of the jet when the throttle is opened.

fuel that can get out into the air stream. The other end of the sliding valve is connected to a cable that goes to the throttle.

A cross-sectional view of a sliding valve carburetor is shown in Figure 9-41. Fuel enters the carburetor and is controlled by a float. Air comes in the side of the carburetor. As the air passes over the venturi, a vacuum is created. Fuel is pulled out of the float bowl through a main jet.

The amount of fuel that can get out of the main jet is determined by the position of the needle. If the rider closes the throttle, the sliding valve moves down and the needle goes way into the main jet. Not much fuel can get out and the engine slows down. As the rider opens the throttle, the sliding valve is pulled up. The needle moves out of the main jet and more fuel can get through, Figure 9-42.

GOVERNORS

We control the speed of an engine with the carburetor throttle valve. If we open the throttle valve, the engine runs fast. When we close the throttle, the engine runs slow. On some engines

Figure 9-43. This mower needs a governor to keep the engine speed steady. (The Toro Co.)

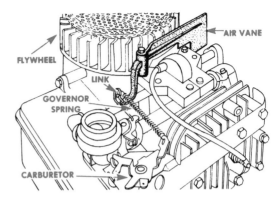

Figure 9-44. Air vane governor parts. (Briggs & Stratton Corp.)

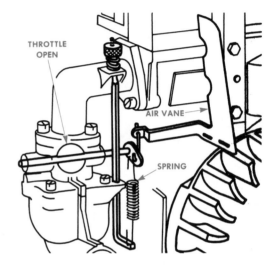

Figure 9-45. The governor spring tries to hold the throttle open. (Briggs & Stratton Corp.)

there is another control over speed. This control comes from a *governor*.

A governor is used to hold an engine at a steady speed. Let's say you are mowing a lawn. You set the throttle on your mower at the right speed. Then you start mowing the grass, Figure 9-43. When you come to very high grass, the mower engine has to work very hard. The engine will start to slow down.

The governor's job is to keep the engine running at a steady speed. It does this by opening or closing the throttle valve. The governor closes the throttle valve if the engine is not working hard. It opens the throttle for more power if the engine has to work hard. Engines used on lawn mowers and tillers use a governor. There are two types of governors. One is called an *air vane* and the other a *mechanical governor*.

Air Vane Governor

An air vane governor uses the force of air coming off the flywheel. A small, flat piece of plastic or steel is mounted above the flywheel, as shown in Figure 9-44. This is called the *air vane*. The air vane is connected to the throttle valve by a piece of wire, called a *link*. A spring is connected to the link. The spring is connected so that it tries to open the throttle, Figure 9-45.

When the engine is running, air is thrown off the flywheel. The air hits the air vane as shown in Figure 9-46. The vane and link try to close the carburetor throttle. The faster the engine is running the more air hits the vane. The stronger the push on the vane, the more force there is to close the throttle valve.

The governor spring works against the air vane. It tries to pull the throttle open. The spring usually is connected to a hand control as shown in

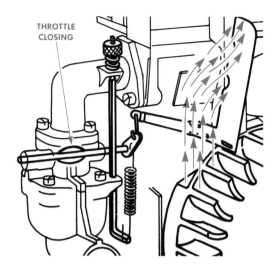

Figure 9-46. When air hits the air vane, the throttle is closed.

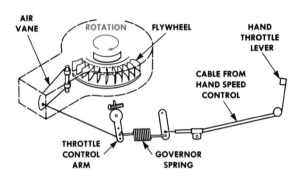

Figure 9-47. The parts of a hand throttle control.

Figure 9-47. When you work the hand throttle lever on a mower, you are pulling on a cable. The cable is connected to the governor spring. Putting tension on the spring will make the engine run faster. Releasing the spring will allow the engine to run slower. The spring and air vane work together to keep the engine speed steady.

Mechanical Governor

Some engines use a mechanical governor. Part of this governor is inside the engine. A small gear rides on the camshaft gear. Attached to the gear is a small set of weights as shown in Figure 9-48. A governor spring, like the one in an air vane governor, is used. The spring holds the throttle open. The spring is attached to the hand control.

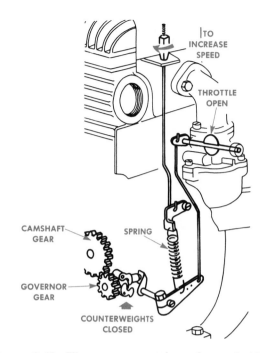

Figure 9-48. The governor spring tries to hold the throttle open. (Briggs & Stratton Corp.)

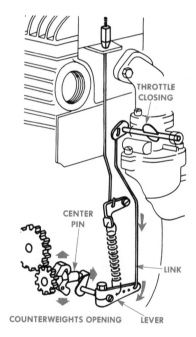

Figure 9-49. As the weights fly out, they move a lever to close the throttle. (Briggs & Stratton Corp.)

As the engine runs, the camshaft gear makes the governor gear turn. The high speed of the governor gear causes the counterweights to fly outward as shown in Figure 9-49. A pin attached to the weights moves. This pushes against an arm and lever. Movement of the lever pulls on the link. The link is connected to the throttle. The throttle is closed. The engine slows down. The spring and counterweights work together to keep the engine speed steady.

AIR CLEANER

On each intake stroke the engine pulls in a lot of air. There is dirt in the air. This dirt must not get into the engine cylinder. Dirt will scratch the cylinder and rings. A dirty engine will wear out very fast. An air cleaner is used to clean the air before it enters the engine.

Oil Bath Air Cleaner

Several types of air cleaners are used on small engines. One common type is called an *oil bath air cleaner*. Like other air cleaners, the oil bath type is mounted to the carburetor. An oil bath air cleaner is shown in Figure 9-50.

All the air going into the carburetor must go through the air cleaner first. The oil bath air cleaner is a can with air passages in it. Dirty air comes in the top of the can as shown in Figure 9-51. The air goes down the inside of the can. The

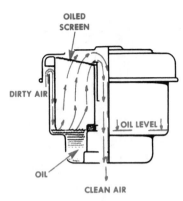

Figure 9-51. A cutaway view of an oil bath air cleaner. (Briggs & Stratton Corp.)

air must then change direction and start back up the can again.

There is a pool of oil in the bottom of the can. The air changes direction right over the oil. Dirt in the air is too heavy to make the quick turn. The dirt keeps going straight into the oil. The oil traps the dirt. The cleaned air goes through an oiled screen. Any dirt left in the air sticks to the oiled screen. Cleaned air then goes down the center of the can and into the carburetor.

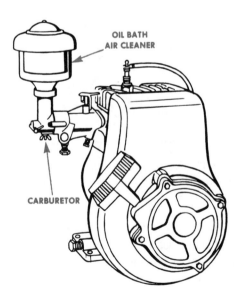

Figure 9-50. An oil bath air cleaner mounted on top of a carburetor. (Clinton Engines Corp.)

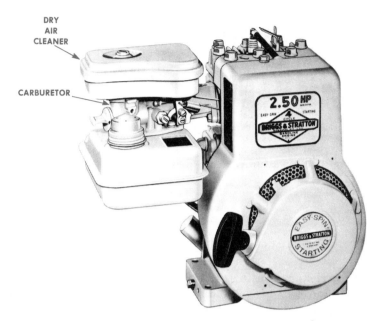

Figure 9-52. A dry air cleaner mounted to a carburetor. (Briggs & Stratton Corp.)

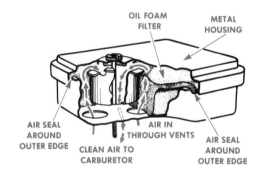

Figure 9-53. A cutaway of an oil foam air cleaner. (Briggs & Stratton Corp.)

Dry Element Air Cleaner

Many small engines use an air cleaner that does not have oil in it. This is called a *dry element air cleaner*, Figure 9-52. One common type of dry element is the foam type, which has a filter element that looks like a sponge. Air enters the foam air cleaner through large holes in the bottom of the metal housing. Air must go through the small holes in the foam filter to get into the engine. The small holes in the foam filter out dirt before it can get into the engine. Air flow through a foam filter is shown in Figure 9-53.

Some dry air cleaners use a paper filter, Figure 9-54. The paper filter has tiny holes in it. Air can

Figure 9-54. Air is cleaned as it goes through this paper filter.

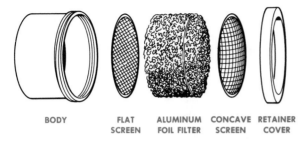

BODY   FLAT SCREEN   ALUMINUM FOIL FILTER   CONCAVE SCREEN   RETAINER COVER

Figure 9-55. Parts of an aluminum foil air filter.

get through the filter paper. Dirt sticks to the outside.

Another type of dry air cleaner uses a filter made of aluminum foil, Figure 9-55. Strips of aluminum foil are pressed together. Air must make its way through the strips. There is a light coat of oil on the aluminum. Dirt sticks to the oil.

NEW TERMS

air cleaner: A filter mounted above the carburetor to clean the air before it enters the engine.

carburetor: Small engine component that mixes fuel with air in the correct amount for combustion.

diaphragm carburetor: A type of carburetor in which the amount of fuel in the carburetor is controlled by a diaphragm.

float: The part of the carburetor that controls the amount of fuel in the float bowl.

float carburetor: A carburetor that controls the amount of fuel with a float.

high-speed adjustment: Carburetor adjustment used to adjust fuel mixture when engine is running at high speed.

low-speed adjustment: Carburetor adjustment used to adjust fuel mixture when engine is running at low speed.

sliding valve carburetor: A carburetor that uses a sliding valve attached to a needle to regulate fuel.

SELF CHECK

1. What engine part mixes the air and fuel?
2. When does the air-fuel mixture go into the engine?
3. How does fuel get into the venturi area on a vacuum or suction carburetor?
4. Explain how a float works to control fuel flow.
5. Describe how a float carburetor works.
6. Why do some engines need a diaphragm carburetor?
7. What is the diaphragm used for?
8. How does a sliding valve carburetor work?
9. What is the purpose of the governor?
10. What is the purpose of the air cleaner?

DISCUSSION TOPICS AND ACTIVITIES

1. Study all the engines in the shop. Can you identify all the different types of carburetors used?
2. Use a cutaway model of any carburetor type. Try to trace the flow of fuel and air through the carburetor.

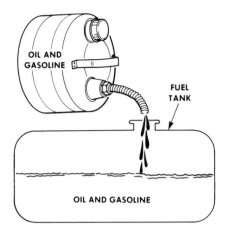

OIL AND GASOLINE

FUEL TANK

OIL AND GASOLINE

unit 10
lubricating systems

An engine needs oil between its moving parts. The oil keeps the parts from rubbing on each other. When the parts do not rub on each other they do not wear out as quickly. The parts also move more easily, because the oil prevents friction. The oil also helps cool the engine by carrying heat away from hot engine parts, and oil is used to clean or flush dirt off engine parts. Oil on the cylinders helps seal the rings to prevent compressed air from leaking. Getting the oil to the engine parts is called *lubrication*. There are several types of lubricating systems used on small engines. In this unit we will study how these systems work.

LET'S FIND OUT: When you finish reading and studying this unit, you should be able to:
1. Describe the purpose of lubrication.
2. Describe the properties of oil.
3. Explain how a two-cycle engine is lubricated.
4. Describe the operation of a splash lubrication system.
5. Explain the operation of a pressure lubrication system.

REDUCING FRICTION

If you push a book along a table top you will notice resistance. This is due to the friction between the book and table. The rougher the table and book surfaces, the more friction there is, because the two surfaces tend to lock together. If a weight is placed on the book, you will notice that it takes even more effort to move it across the table. As the amount of pressure between two objects increases, their friction increases. The type of material from which the two objects are made also affects the friction. If the table is made of glass, the book slides across it easily. If it is made of rubber, it is very difficult to push the book across.

Friction is not desirable between engine parts for several reasons. First, power is needed to overcome friction. The less friction between engine parts, the more usable power an engine can develop. Friction between two objects causes them to heat up and to wear. The fact that friction causes heat may be demonstrated simply by rubbing your hands together rapidly. The heat is caused by the friction between the skin of your two hands.

The purpose of lubrication is to reduce friction on engine parts as much as possible. Friction cannot be eliminated completely, but it can be reduced to a point where long engine life may be expected. Suppose a slippery liquid such as oil is spilled on the table top, Figure 10-1. The book could now be pushed across the table with very little resistance. The friction has been reduced between the book and table. The oil forms a thin layer called a *film* which gets under the book and actually lifts it off the table surface. An oil film is used between engine parts to reduce friction and wear.

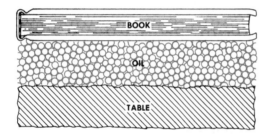

Figure 10-1. Oil reduces the friction between the book and the table.

OIL

Oil is the most common fluid used to provide lubrication. For many years, lubricating oil has been made or refined from crude petroleum pumped from oil wells. Lubricating oil is only one of the many products refined from crude petroleum. Others include gasoline, kerosene and fuel oil. Oil refined from crude petroleum sometimes is called *mineral oil* or *petroleum oil*. Oil also may be made from nonpetroleum sources. This kind of oil is called *synthetic oil*. Oil used in a small engine must have the correct viscosity and service rating for that engine, Figure 10-2.

Figure 10-2. The oil used in an engine must have the correct viscosity and service rating.

Viscosity describes the thickness or thinness of fluid. A common term for viscosity is *body*. A fluid with a high viscosity is said to have a heavy body. An example of fluid with a high viscosity is the thick lubricant used in some standard transmissions and rear axles. A low-viscosity fluid is one with a light body. The viscosity of fluid determines how freely it flows. Low-viscosity fluids flow very freely. High-viscosity fluids flow sluggishly. Oils used in engines must have a viscosity which allows them to flow freely under cold conditions but which has enough body to provide a coating during times of high temperature.

The Society of Automotive Engineers (SAE) has set up standards for oil viscosity. Thin oil receives a low viscosity number, such as SAE 20, while thicker oil receives a higher number, such as SAE 40 or SAE 50. The viscosity number is stamped on top of the oil can. The operator's manual for the engine usually specifies what viscosity should be used.

Some engines used SAE 30 in the summer and SAE 20, 10 or 5 in the winter. For this reason, oil companies sell what is called a multiple-viscosity oil, such as SAE 10-40. This oil flows freely like SAE 10 when the weather is cold, but protects like SAE 40 when it is hot. An engine operated in different climates needs multiple-viscosity oil. A viscosity rating that has a W after it, such as SAE 20W, means the oil has been tested at 0° F and is rated for cold-temperature operation.

Just as important as viscosity is an engine oil's service rating, printed on the side of the oil can. The service ratings, set up by the American Petroleum Institute (API), are a measure of how well the oil holds up under severe wear-and-tear. If the oil does not do well, it gets a low rating or classification; if it does well, it gets a high one. The categories are SA (lowest) SB, SC, SD, SE (highest).

Small engines may use oil of a lower classification than automobiles. The engine operator's manual usually will specify what service classification should be used. A higher classification can always be substituted but never a lower one.

LUBRICATION SYSTEMS

Premix Lubrication

Most two-stroke engines are lubricated by premix lubrication. Both oil and gasoline go in the fuel tank, as shown in Figure 10-3. Usually they are mixed together in the right amount and then the mixture is poured into the fuel tank.

The oil is a special type of two-stroke-cycle oil, Figure 10-4. It mixes with the gasoline and stays mixed for a long time. The mixture is different for different engines. Some engines use 20 parts of gasoline for each part of oil. Other engines use a 50 to 1 mix. This means 50 parts of gasoline are used with one part of oil. There is always more gasoline than oil in the mixture. It is possible to

Figure 10-4. Two-stroke oil is made to be mixed with gasoline.

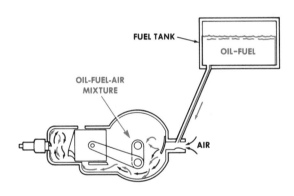

Figure 10-5. A mixture of oil and fuel goes into the crankcase with the air.

buy fuel with oil already mixed in. This is called *premix.*

The oil-and-fuel mixture goes into the engine's crankcase through the reed valve, as shown in Figure 10-5. The oil in the mixture floats around

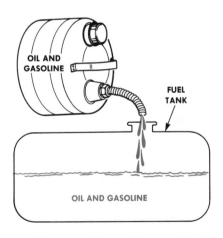

Figure 10-3. Fuel and oil are mixed together in the fuel tank of a two-cycle engine.

in the crankcase as a mist. Some of the oil falls on the engine parts. This is the oil that lubricates. Oil droplets fall on the crankshaft connecting rod, main bearings, piston pin and cylinder walls. Some of the oil also goes with the fuel into the combustion chamber, where the oil is burned along with the fuel.

The main advantage of the premix lubrication system is that it has no moving parts. On the other hand, if the operator forgets to mix oil with the gasoline, the engine will be damaged.

Oil Injection

There is another way to get the fuel and oil mixed together. Many larger two-stroke-cycle motorcycle engines have two tanks. One of the tanks, the fuel tank, holds gasoline. The other, the oil tank, holds oil. A pump is used to pull oil out of the oil tank. The oil moves up an oil line to the intake port and is forced out a small nozzle called a *port injector* into the air-fuel mixture. The oil mist then is carried around the crankcase. It settles on the parts and provides the lubrication. A port injection system is shown in Figure 10-6.

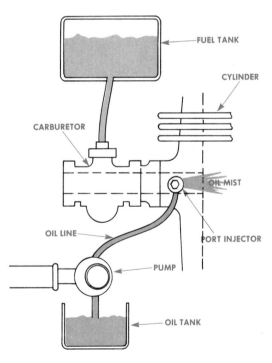

Figure 10-6. Oil is pumped into the port in a port injection system.

Splash Lubrication

Most of the small four-stroke-cycle engines used on lawn equipment use a splash lubrication system. Splash lubrication needs very few parts. Horizontal-crankshaft engines use an oil dipper. The dipper is hooked to the bottom of the connecting rod. An oil dipper is shown in Figure 10-7. When the rod goes around, the dipper goes around. The dipper is used to splash oil on the engine parts.

Vertical-crankshaft engines cannot use a dipper, because the connecting rod is not near the oil. A small gear called a *slinger* is used. The gear rides on the camshaft gear, as shown in Figure 10-8.

Figure 10-7. Horizontal crankshaft engines use a dipper on the connecting rod.

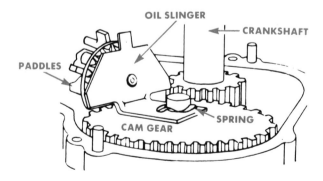

Figure 10-8. Vertical crankshaft engines use an oil slinger driven by the camshaft.

When the camshaft turns, the slinger turns. The small paddles on the slinger dip into the oil. The oil is splashed on the engine parts.

The engine's crankcase is partly filled with oil, as shown in Figure 10-9. As the engine runs, the crankshaft and connecting rod turn around and around. The oil dipper on the connecting rod dips into the oil, Figure 10-10, and splashes the oil upward.

Some of the oil is splashed on the cylinder. It is carried upward by the piston and rings. This provides lubrication between the piston and cylinder wall.

The oil that is splashed up by the dipper falls back down and into holes in some of the engine parts designed to catch it. There is an oil hole in the connecting rod, as shown in Figure 10-11. Oil runs into this hole. The hole leads into the area where the rod is attached to the crankshaft. Oil gets between the connecting rod and crankshaft.

There are also holes or passages in each of the main bearings. Oil runs into the main bearing

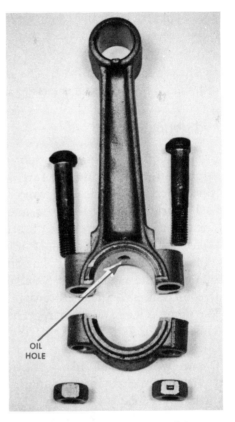

Figure 10-11. An oil hole in the side of the connecting rod lets oil get between the rod and crankshaft.

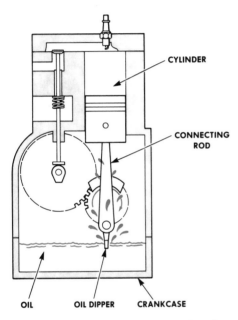

Figure 10-9. The oil dipper splashes the oil up on the engine parts.

Figure 10-10. A dipper going into the oil.

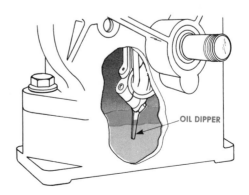

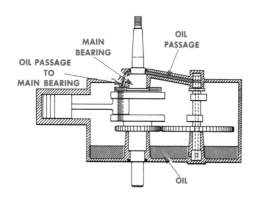

Figure 10-12. Oil runs down through holes and into the main bearings. (Kohler)

holes, as shown in Figure 10-12. The oil gets between the main bearings and the crankshaft.

Pressure Lubrication

Some small engines use pressure lubrication. Pressure lubrication uses a pump to force oil into engine parts. Pressure lubrication works better than splash lubrication. Oil is forced into parts such as the connecting rod and main bearings.

Oil Pump. The oil pump is the most important part used for pressure lubrication. The oil pump makes oil flow to the engine parts. Small engines are a barrel oil pump like the one shown in Figure 10-13. The pump is made in two parts. The larger part is called the *body*. A smaller part, called a *plunger*, fits inside.

The pump works much like a piston and cylinder. In principle, when the plunger moves out of the body, it creates a vacuum. This movement is called an *intake stroke*. When the plunger moves into the body, it is called a *compression stroke*. The pump pulls in oil on the intake stroke and then pushes oil into engine parts on the compression stroke. The two strokes of a pump are shown in Figure 10-14.

The pump body has a large hole in it. This hole

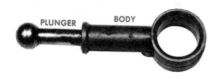

Figure 10-13. A barrel pump has a body and plunger.

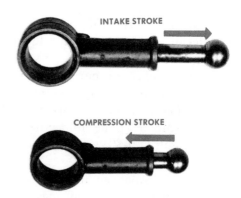

Figure 10-14. The pump pulls oil in on the intake stroke and forces it into engine parts on the compression stroke.

is drilled to fit around a lobe on the camshaft. The plunger is mounted so that it does not move.

When the engine is running, the camshaft goes around and around. The offset hole causes the body to move up and down. The pump plunger cannot move. The body slides in and out over the plunger. As the pump has an intake stroke, oil is pulled into it. When it has a compression stroke, oil is forced into engine parts.

Oil Flow. Oil holes are used to get the oil into the engine parts. The pump pulls oil out of the crankcase on the intake stroke. Oil is pushed out of the pump on its compression stroke. The oil leaves the pump through an oil hole, as shown in Figure 10-15.

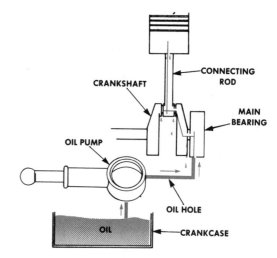

Figure 10-15. The pump forces oil into the engine parts.

The oil goes through a hole into the main bearings. Here it gets between the main bearing and crankshaft. This gives the main bearings their lubrication.

There is another hole drilled in the crankshaft. This hole lets oil flow from the main bearing to the connecting rod bearing. Oil can get between the crankshaft and the connecting rod.

Some engines have a hole that goes up the middle of the connecting rod. Oil can flow up this hole. This oil flows in between the connecting rod and piston pin.

When the engine is running, oil leaks out between the crankshaft and connecting rod. This oil is thrown off the moving crankshaft. Some of this oil is thrown up on the cylinder wall. This is how the cylinders get lubricated.

Dry Sump Lubrication

Some motorcycles use a dry sump lubrication system. This system uses a pump to force oil through the engine parts and a pump to return the oil to the oil tank. The parts of a dry sump system are shown in Figure 10-16. The oil tank usually is located above the engine. Oil flows by gravity into the high-pressure pump. The pump forces oil through the engine parts for lubrication. The lubricating oil runs down into the bottom of the engine. The area where the oil is collected is called a *sump*. A scavenger pump pulls the oil out of the sump and returns it to the oil tank. Since there is never very much oil in the sump, the system is called a *dry sump*. The advantage of this system is

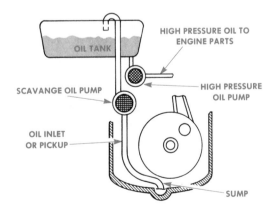

Figure 10-16. A dry sump lubricating system.

that more oil is available for the engine in the tank than could be stored in the engine itself.

NEW TERMS

dry sump lubrication: A lubrication system that uses a scavenger pump to pull oil out of the engine into an oil tank.

friction: Resistance to motion between two parts that causes wear and heat.

lubrication: Reducing friction in an engine as much as possible by providing oil between moving parts.

oil: A fluid used for lubrication.

premix lubrication: Oil and gasoline mixed together for lubrication.

pressure lubrication: A system that uses a pump to force oil into engine parts.

service classification: A rating system for engine oil based on the type of service it is used in.

splash lubrication: A system in which oil is splashed on engine parts for lubrication.

viscosity: The thickness or thinness of the oil.

SELF CHECK

1. Why is friction bad in an engine?
2. Describe how oil can reduce friction.
3. How can you find out the viscosity of an oil?
4. How can you find out the service rating of an oil?
5. Describe how a premix lubrication system works.
6. What is oil injection?
7. Explain how a splash lubrication system works.
8. Explain how a pressure lubrication system works.
9. What is the purpose of the oil pump in a pressure lubrication system?
10. Why does a dry sump lubrication system need two pumps?

DISCUSSION TOPICS AND ACTIVITIES

1. Look up the correct oil viscosity and service classification for a small engine.
2. Use a cutaway model to trace the oil flow in a pressure lubrication system.

unit 11
cooling systems

During the engine's power stroke, a mixture of air and fuel is burned in the cylinder. This burning creates a lot of heat. Much of this heat is used to push down the piston. Some of the heat goes into the engine's parts. We need a way to take the heat away from these engine parts; otherwise the parts could be damaged. In this unit we will see how engines are cooled.

LET'S FIND OUT: **When you finish reading and studying this unit, you should be able to:**
   1. **Describe the purpose of the engine cooling system.**
   2. **Describe the operation of the draft-type air-cooling system.**
   3. **Explain the operation of the forced-air-circulation cooling system.**
   4. **List the components used in a liquid-cooling system.**
   5. **Trace the flow of coolant through a liquid-cooling system.**

DRAFT AIR COOLING

There are several ways in which engines are cooled. One way is called *liquid cooling*. A liquid such as water circulates around all the hot engine parts. The water takes away the heat. Most automobile and outboard engines are cooled this way.

Many small engines are air-cooled. Air goes around the engine parts and takes away the heat. Air may be forced around the engine parts by a natural draft or by forced circulation. Most motorcycles and mopeds use a draft cooling system.

The components that get the hottest, such as the cylinder and cylinder head, have fins, Figure 11-1, to direct the greatest amount of air into contact with the greatest amount of hot metal. When the engine is running, heat builds up in the cylinder head and cylinder. As the heat goes through the cylinder head and cylinder, it moves out into the cooling fins, as shown in Figure 11-2.

Heat must be removed from the cooling fins. This may be done with a natural draft. As a motorcycle, Figure 11-3, or moped moves along, it pushes through the air. This air flows over the cooling fins and carries away the heat. The faster

Figure 11-1. Air-cooled engines have cooling fins on cylinders and cylinder heads. (B.M.W. of North America Inc.)

Figure 11-3. Air moves around the engine parts to cool them as this motorcycle goes down the road. (B.M.W. of North America Inc.)

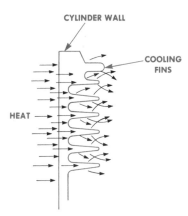

Figure 11-2. Heat goes through the engine parts and into the cooling fins.

the motorcycle goes, the more air flows over the engine. The only difficulty with this system is that when the motorcycle is stopped and the engine is

running, there is no air flow over the engine. If the engine runs too long under these conditions, it will overheat.

FORCED-AIR COOLING

Many small engines used on lawn equipment and stationary equipment do not move or do not move fast enough for draft cooling. Air has to be forced around the parts with a pump. In this section will see how the forced-air cooling system works.

The main air-cooling parts are cooling fins, flywheel fins and a blower housing. The engine parts that get the hottest are the cylinder and cylinder head. Both of these parts have cooling

COOLING FINS

BLOWER HOUSING

FLYWHEEL FINS

Figure 11-4. Forced-air cooling parts.

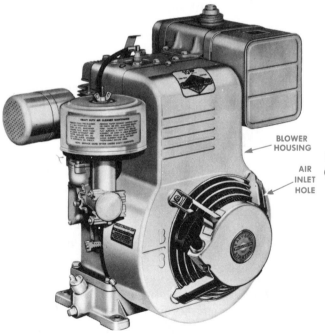

BLOWER
HOUSING

AIR
INLET
HOLE

Figure 11-5. The blower housing covers the flywheel. (Briggs & Stratton Corp.)

fins, similar to those on a motorcycle or moped engine, to control heat and air flow.

The flywheel is used as a pump to move air around the engine parts. The flywheel has fins that help it pump the air. An engine with cooling fins and flywheel fins is shown in Figure 11-4.

The flywheel is covered over with a blower housing. The blower housing is used to direct the cooling air. There is an air inlet hole in the middle of the blower housing. Cooling air comes in this hole. A blower housing with an inlet hole is shown in Figure 11-5.

The flywheel turns when the engine is running. The fins on the flywheel make it work like a fan. Cool air is pulled in the inlet hole of the blower housing. The inlet hole has a screen covering it, Figure 11-6. The screen keeps grass and other foreign objects out of the housing.

The shape of the blower housing controls where the air goes. As shown in Figure 11-7, air goes up to the top of the blower housing. It then goes through each of the cooling fins. The cool air takes heat away from the fins, and the engine parts are cooled down, Figure 11-8.

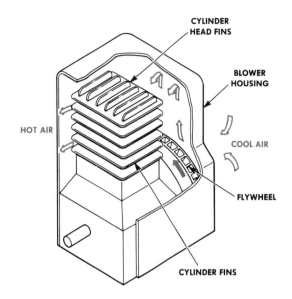

Figure 11-7. Cool air is pumped by the flywheel around the hot engine parts to cool them.

Figure 11-8. Where air goes in and comes out of an engine. (Briggs & Stratton Corp.)

LIQUID COOLING

Most outboard engines and a few motorcycle engines have a liquid cooling system. The liquid cooling system circulates liquid around hot

Figure 11-6. A screen is used to keep grass and leaves from going into the blower housing. (Briggs & Stratton Corp.)

engine parts to carry off the heat. Coolant passages called *water jackets* surround each cylinder in the block and the cylinder head very close to the valve area. Heat from the burning air-fuel mixture passes through the metal of the cylinder head and cylinder wall and enters the water jackets. The heat then goes into the liquid coolant circulating through the water jackets.

An advantage of the liquid cooling system is that it can take care of more heat than can air-cooling. A liquid carriers heat more efficiently than a gas (air) does. A multi-cylinder liquid-cooled engine is also less expensive to make, since the cylinders may be cast together in a block instead of made separately. An additional advantage is that liquid-cooling passages reduce engine noise so that engine operation is much quieter.

OUTBOARD COOLING SYSTEMS

Outboard engines use liquid cooling because often they are large, multi-cylinder engines which would be difficult to cool by air. Since they are always operated in water, they have a ready source of coolant. Cooling water is pulled into the lower unit of an outboard near the propeller. The water is directed up into the power unit, where it circulates through passages in the cylinder and cylinder head. The water removes the heat. The heated water is directed back down to the lower unit where it goes out of the engine. The water circulation through an outboard is shown in Figure 11-9. Water is pulled into the outboard and circulated through the engine by a water pump. The pump, usually located in the lower unit, is driven by the drive shaft that drives the propeller. The pump consists of a synthetic rubber impeller, which is keyed to the driveshaft, and a pump housing, which is offset from center with respect to the driveshaft. As the impeller spins, a low-pressure area is created in its center. Water is drawn into the center of the impeller and is thrown off the impeller blades by centrifugal force. This causes water to be drawn in and pushed out of the pump.

Many outboards use a pump that has a variable volume. It pumps less water at high speed. This is necessary because the pump turns so fast at high

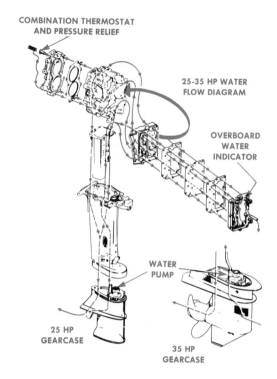

Figure 11-9. Coolant flow through an outboard engine. (Evinrude Motors)

speed that water would move too fast through the engine. Also, engine power is needed to turn the pump. If the pump does not have to work as hard at high speed, less engine power is wasted. This is achieved by making the pump impeller blades flexible. At low speeds the blades contact the housing, creating a good seal between the blades. A high volume of water is pulled in. At high speed, water resistance prevents the blades from touching the housing. The pump is less effective and the volume of water decreases. The shapes of the impeller blades at high and low speed are shown in Figure 11-10. An outboard engine must

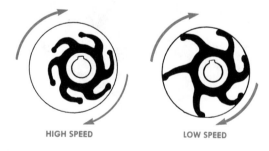

HIGH SPEED   LOW SPEED

Figure 11-10. Shape of impeller blades at high and low speed. (Evinrude Motors)

always be operated in water so that there is a coolant flow. In the shop, outboards are operated in a water tank like the one shown in Figure 11-11.

Efficient temperature control is achieved in a liquid-cooling system by regulating the flow of coolant through the system with a thermostat, Figure 11-12. The thermostat is a temperature-controlled valve which controls the flow of water into the radiator from the engine.

A thermostat prevents overcooling. An engine operating at too low a temperature becomes less efficient. During the power stroke, heat from the burning mixture is pushed down the piston. If too much of this heat is lost to the cooling system,

Figure 11-11. Outboards must be operated in a water tank in the shop for cooling.

Figure 11-12. A thermostat regulates the flow of coolant.

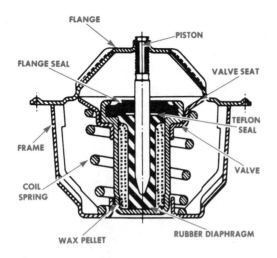

Figure 11-13. Sectional view of a thermostat. (Chevrolet Div. — General Motors Corp.)

power and efficiency also are lost. If the cylinder area is too cool, fuel will not burn completely. Some of the fuel may run down the cylinder walls past the rings, washing off lubricating oil. Enough gasoline may enter the oil pan to dilute the oil. During the exhaust stroke, unburned gasoline that did not wash down the cylinder is pushed out, adding to exhaust emissions. Whenever the fuel is not burned completely, power is lost and fuel economy suffers.

In a pellet thermostat, Figure 11-13, a wax pellet or power element in the thermostat expands when heated and shrinks when cooled. The pellet is connected through a piston to a valve. The heated pellet pushes against a rubber diaphragm which forces the valve to open. As the pellet shrinks on cooling, it allows a spring to close the valve and stop circulation of coolant through the power head.

As the engine becomes warm, the pellet gets big and the thermostat valve opens, permitting the coolant to flow through the radiator. This opening and closing of the thermostat valve permits enough coolant to enter the power head to keep the engine within operating-temperature limits. The thermostat is mounted in the coolant passage leading to the power head.

MOTORCYCLE COOLING SYSTEMS

As mentioned previously, most motorcycles have air-cooled engines. A few large touring motorcycles use a liquid-cooling system. A liquid

is circulated through the engine to carry away the heat. A coolant pump circulates the coolant and a thermostat regulates the flow. These parts work like those described for outboards.

The heat removed from the hot engine parts by the coolant must then be removed from the coolant. This is done by pumping the hot coolant out

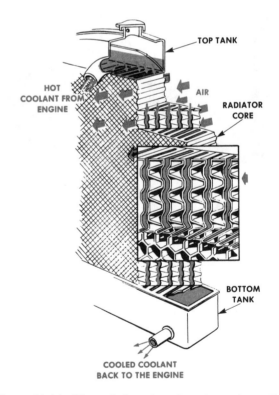

Figure 11-14. Flow of air and coolant through a radiator. (E. I. DU Pont DE Nemours & Co.)

of the engine and into a heat exhanger, commonly referred to as a *radiator*. The radiator removes heat from the coolant.

The radiator, mounted in front of the engine, is made up of a top tank, a bottom tank and a center core or heat exchanger. Hot coolant is pumped out of the engine through a large hose connected to the top tank. It enters the radiator core through several small distribution tubes. These tubes are made from a metal that is a good heat conductor, usually copper or aluminum. The heat passes out of the liquid and into the wall of the tubes, which are fitted with copper or aluminum fins. Air circulated through the core by the motorcycle moving through the air takes the heat from the fins. The cooled liquid runs into the bottom tank of the radiator. A large hose allows the coolant to be drawn from the bottom tank back into the engine to pick up more heat, Figure 11-14.

NEW TERMS

air cooling: Cooling engine parts by circulating air around them.

air pump: Pump used with an air-cooling system to force air around hot parts.

coolant: Liquid used in liquid-cooling system to carry away heat; usually a mixture of ethylene glycol and water.

coolant pump: Pump used to circulate coolant around hot engine parts.

cooling fins: Metal fins used on air-cooled engine parts to move heat away from the parts.

cooling system: An engine system used to keep the engine's temperature within limits.

radiator: A large heat exchanger located in front of the engine.

thermostat: A device in the cooling system used to control the flow of coolant.

water jackets: Passages in the cylinder block and head for coolant flow.

SELF CHECK

1. Why do engine parts get hot?
2. What can happen if engine parts get too hot?
3. What are the three ways in which engines are cooled?
4. How is an air draft used to cool an engine?
5. List the main forced-air cooling parts.
6. What are cooling fins used for?
7. What does the flywheel do in air cooling?
8. How does liquid cooling work?
9. Explain the purpose and operation of the coolant pump.
10. Describe the purpose of the radiator.

DISCUSSION TOPICS AND ACTIVITIES

1. Use a cutaway model of an engine to trace the flow of cooling air through the cooling system.
2. Put a thermostat in a pan of hot water and observe opening of the valve.

We will study the operation of
trail bikes. (Kawasaki Motors
Corp., Inc.)

122

part 4
small engine performance

Small engines are measured and compared by size and performance measurements. We describe engines by how many cubic inches of displacement they have or how much horsepower they develop. In this part we will see how these measurements are determined and what they mean. You will be able to compare one engine with another.

unit 12

size and performance measurement

Small engines are described and compared by a number of measurements related to size and performance. Measurements of engine size are concerned with an engine's bore, stroke displacement and compression ratio. These measurements determine how much power an engine can develop. The amount of power an engine actually develops is specified in terms of performance measurement called *horsepower*. The concept of horsepower is based upon a number of scientific principles such as force, work, power, energy and efficiency. Each of these principles is explained below.

LET'S FIND OUT: When you finish reading and studying this unit, you should be able to:
1. Understand the size measurements based on bore, stroke, displacement and compression ratio.
2. Identify the elements of small-engine performance measurement for linear horsepower.
3. Recognize the elements of engine rotary horsepower measurement.
4. Describe the way horsepower is measured, charted and rated.
5. Explain the different types of efficiency ratings used with small engines.

ENGINE SIZE MEASUREMENTS

Engines are manufactured in different sizes to meet different requirements. A lawnmower may have a small engine and an outboard a large one. Engine size comparisons are not based on the outside dimensions of an engine but on the size of the area where power is developed.

Bore and Stroke

The bore is the diameter of the cylinder, Figure 12-1. The larger the bore, the more powerful the engine. The stroke is the distance the piston moves from the bottom of the cylinder to the top or from the top to the bottom. The size of the stroke is determined by the distance between the

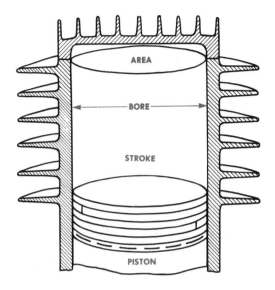

Figure 12-1. The bore is the distance across the cylinder. (Briggs & Stratton Corp.)

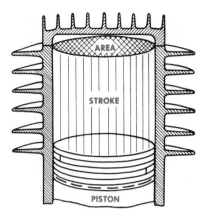

Figure 12-2. Displacement is the volume of the cylinder when the piston is at the bottom of its stroke. (Briggs & Stratton Corp.)

centerline of the crankshaft and the centerline of the connecting rod where it attaches to the crankshaft. The longer the stroke, generally speaking, the more powerful the engine. The bore and stroke are given in inches or millimeters, e.g., a bore of 3½ inches and a stroke of 4 inches, or a bore of 84 mm and a stroke of 88mm. Sometimes just the numbers are given: 3½ × 4, for example. In this case, the bore is always the first number.

Displacement

The bore and stroke of an engine are used to find its displacement. Displacement is the size or volume of the cylinder, Figure 12-2. Displacement is measured when the piston is at the bottom of the cylinder. The bigger the bore, the larger the cylinder volume or displacement. The longer the stroke, the larger the cylinder volume or displacement. If the engine has more than one cylinder, the displacement of all the cylinders is added together. This gives us the total displacement for the engine.

The displacement of an engine may be given in two ways. It may be measured in cubic inches or in cubic centimeters. We often just call cubic centimeters cc's. Usually the larger an engine's dis-

placement, the more powerful the engine is. The displacement of many small engines is 6 to 8 cubic inches. A car engine may have a displacement of more than 200 cubic inches.

The formula used to calculate displacement is:

$$\text{Displacement} = \frac{Bore^2 \times \pi \times \text{Stroke}}{4}$$

The bore is squared, or multiplied by itself. The symbol π stands for 3.14. An engine with a bore of 3 inches and a stroke of 3 inches would have a displacement calculated as follows:

$$\text{Displacement} = \frac{3^2 \times 3.14 \times 3}{4} =$$

$$\frac{9 \times 3.14 \times 3}{4} =$$

21.20 cubic inches

The displacement in one cylinder is multiplied by the number of cylinders to find the displacement for the entire engine. If the engine in the example above has two cylinders, the displacement is 42.4 cubic inches.

Compression Ratio

An engine's compression ratio is another important engine measurement.

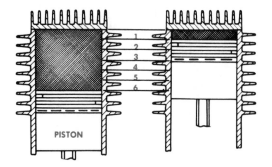

Figure 12-3. This engine has a compression ratio of 6 to 1. (Briggs & Stratton Corp.)

You may remember that one of the four strokes of the four-stroke-cycle engine is called a *compression stroke*. On this stroke the piston moves up. Both the intake and exhaust valve are closed. The piston squeezes the air-fuel mixture in the top of the cylinder. The tighter the mixture is squeezed, the higher the compression an engine has. The higher the compression, the more power the engine can develop.

We need a way to measure how much compression an engine has. The measurement we use is called *compression ratio*. Compression ratio is measurement of how tightly the air-fuel mixture is squeezed in the cylinder. We find the compression ratio by measuring the area of a cylinder. First we measure the area when the piston is at the bottom of its stroke. The piston is then moved to the top of its stroke. The small area above the piston is measured.

Let's see how this works. In Figure 12-3, the piston on the left is at the bottom of its stroke. If we measure the area we find there are six cubic inches. The piston is moved to the top of its stroke (right). We find there is only one cubic inch of space now. This means we squeeze six inches of air-fuel mixture into a one-cubic-inch area. We call this a compression ratio of 6 to 1.

ENGINE PERFORMANCE MEASUREMENTS

More than two hundred years ago an engineer and practical instrument maker at Glasgow University was conducting experiments to improve the steam engine. The experiments were so successful that a practical steam engine was devel-

oped. The engineer, James Watt, went on to form a company to sell the engines. The engines were designed to be used in the coal mines in England. Up to this time, the mine operators did all the lifting of coal with horses. To sell his engines, Watt had to develop a system for comparing his engines to the horse. The biggest achievement of James Watt was to establish a unit for power. He thought up the term *horsepower* after establishing by experiment the amount of work an average horse could do.

The idea of horsepower is based upon a number of simple scientific principles: force, work, power, torque, energy and efficiency. Each of these important principles and its relationship to horsepower is discussed below.

Force

In simple terms, a force is a push or pull. If a man pushes against a door, he has exerted a force on the door. In scientific terms, a force is explained as an action against an object that tends either to move the object from a state of rest or to change the direction or speed of an object already in motion. The amount of force exerted can, of course, be measured. In the customary system, force is measured in pounds. To open a door, a person must push against the door with a force of so many pounds. In the metric system, force is measured in newtons.

Work

The term *work* is familiar to everyone. In scientific terms, however, work is done when a force travels through a distance. Work is done only if

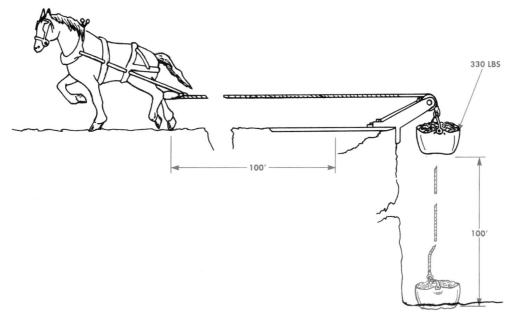

Figure 12-4. Work is force times distance.

the force results in movement. Someone who pushes against a door that is locked exerts a force. If the door is not moved, no work is done.

James Watt was interested in the amount of work a horse could do. He hitched a horse to a container of coal that weighed 330 pounds. He had the horse pull the container of coal 100 feet Figure 12-4. The formula for work is:

Work = Distance × Force

Using the example of the horse:

Work = distance (feet) × force (pounds)

Work = 100 feet × 330 pounds

Work = 33,000 foot-pounds

The horse, then, accomplished 33,000 foot-pounds of work. In the United States, the measurement of work is the foot-pound. One foot-pound equals the force of one pound moved through a distance of one foot. In the metric system, work is measured in joules. One joule is equal to a force of one newton moved through a distance of one meter. These ideas can be expressed as an equation:

Work = Distance × Force

foot-pounds = feet × pounds

joules = meters × newtons

Power

Power is the rate or speed at which work is done. Power adds the idea of time. The faster work is done, the more power is involved. Consider again the example of the horse. If the horse is able to pull the container of coal one hundred feet in one minute, two horses might be able to pull the container of coal one hundred feet in half a minute, Figure 12-5. The same amount of work is done in both cases, but the amount of power involved is different. The formula for power is:

$$\text{Power} = \frac{\text{work}}{\text{time}}$$

$$\text{Power} = \frac{\text{distance} \times \text{force}}{\text{time (minutes)}}$$

$$\text{Power} = \frac{100 \text{ feet} \times 330 \text{ pounds}}{1 \text{ minute}}$$

$$\text{Power} = \frac{33,000}{1}$$

Power = 33,000 foot-pounds per minute

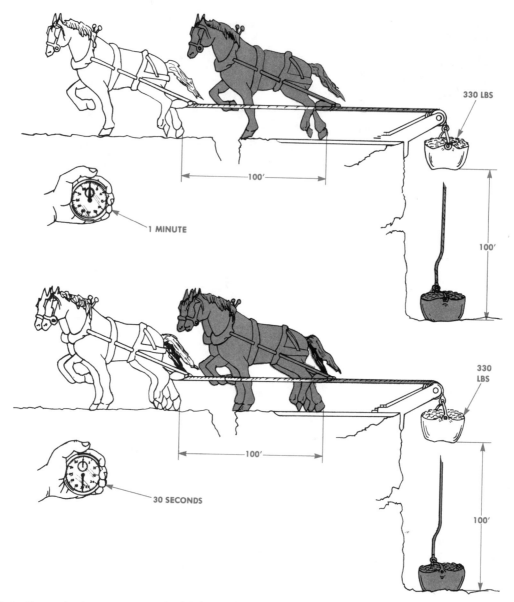

Figure 12-5. Power is the rate or speed of doing work.

When two horses pull the same weight 100 feet in 30 seconds or .5 minute, the amount of power is:

$$Power = \frac{work}{time}$$

$$Power = \frac{distance \times force}{time\ (minutes)}$$

$$Power = \frac{100\ feet \times 330\ pounds}{.5\ (minute)}$$

$$Power = \frac{33,000}{.5}$$

Power = 66,000 foot-pounds per mintue

From this example, it can be seen that the faster work is done, the more power is developed. In the customary system of measurement, power commonly is described in foot-pounds per minute. In the metric system, the unit of power measurement is the watt. One watt is equal to one joule per second. A joule is equal to a newton moved through one meter.

Horsepower. The steps used above to determine the amount of power a horse could develop are like those James Watt used. After watching the power produced by draft horses, Watt decided that 33,000 foot-pounds per minute was about what the average horse could do. A horsepower is the ability to do 33,000 foot-pounds of work in one minute. He then was able to write a formula for horsepower. The formula became:

$$\text{Horsepower} = \frac{\text{Distance (feet)} \times \text{Force (pounds)}}{\text{Time (minutes)} \times 33,000}$$

Again, using the example of the horse:

$$\text{Horsepower} = \frac{100 \text{ feet} \times 330 \text{ pounds}}{1 \text{ minute} \times 33,000}$$

$$\text{Horsepower} = \frac{33,000}{33,000}$$

Horsepower = 1

This idea of horsepower allowed James Watt to compare the steam engines he was trying to sell to the common power-producer of his day — the horse. This proved to be a very useful idea, but the formula was limited to power used for pulling a weight a straight distance. The rotating crankshaft of an engine, however, does not develop a pulling force. Instead, it develops a force through a circle as it turns.

Rotary Horsepower (Torque). To measure power developed by an engine crankshaft, a rotary unit of force is necessary. The rotary unit of force is called *torque.* Torque, in simple terms, is turning or twisting effort. A mechanic using a wrench to tighten a bolt is applying torque to the bolt. When the bolt is tight, the mechanic may not be able to turn it any more. Even though the bolt does not turn, the mechanic is applying torque. Torque, then, is a force that produces or tries to produce rotation. If the torque results in rotation, work is done.

The formula for determining torque is:

Torque = Force × Radius

The term *force* is used exactly as it is in the concept of work. The radius is the distance from the

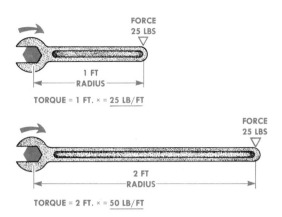

Figure 12-6. The longer the radius the more torque you can get with the same force.

point at which the force acts to the center of rotation of the shaft. The push or pull the mechanic exerts on the wrench is the force. The distance from the center of the bolt to the part of the wrench handle where the mechanic applies the force is the radius. It is called a radius because as the wrench goes around it describes a circle. The wrench takes up one-half the diameter of the circle, or the radius.

If the wrench the mechanic is using, Figure 12-6, has a 1-foot radius and the force exerted is 25 pounds, the torque may be calculated as follows:

Torque = Force × Radius
Torque = 25 lbs × 1 ft
Torque = 25 lbs ft

Should the mechanic choose a longer wrench with a 2-foot radius and exert the same amount of force, the torque will be increased.

Torque = Force × Radius
Torque = 25 lbs × 2 ft
Torque = 50 lbs ft

In the customary system, torque always is expressed in pounds-feet. This causes a good deal of confusion because the unit of measurement for work is foot-pounds. In both formulas, pounds are multiplied by feet. Remember that torque always is specified in pounds-feet and work in foot-pounds. In the metric system, torque is measured in newton-meters. The formulas are:

Torque = Force × Radius
Customary system: Torque = Pounds × Feet
Metric system: Torque = Newtons × Meters

The power produced by an engine is rotary. Like linear horsepower can be measured. Rotary horsepower measurement is based upon the formula Watt developed. The formula for rotary horsepower is:

$$\text{Horsepower} = \frac{\text{RPM} \times \text{Torque}}{5252}$$

Torque is the turning effort developed by the engine. RPM means revolutions per minute. RPM gives us the time element in the formula. The 5252 is simply a number called a constant by which we divide to get the correct units.

MEASURING HORSEPOWER

The horsepower of an engine can be found if the engine's torque at any particular RPM is measured. A dynamometer, Figure 12-7, is a device used to measure engine torque. The dynamometer does not measure horsepower directly. It measures torque and RPM. These values are then put into the horsepower formula and horsepower is determined mathematically. Many dynamometers have the ability to do this math automatically and provide the operator with a horsepower figure.

There are many types of dynamometers. Most dynamometers measure torque of an engine by

Figure 12-7. A small engine dynamometer.

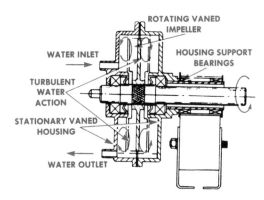

Figure 12-8. Cross section of a water brake unit. (Go-Power Corp.)

changing the rotating torque to a stationary torque. The stationary torque is then measured with a scale, a hanging weight, load cell strain gage or other force-measuring device at the end of a torque arm.

Most dynamometers use a hydraulic water brake to change rotating torque to stationary torque and absorb the power. The water brake absorption unit shown in Figure 12-8 consists of a vaned impeller that rotates in a stationary vaned housing. When the absorption unit is partly filled with water, the vaned impeller rotates and accelerates the water outward in the direction of rotation of the impeller until the water strikes the outer edge of the housing. The water is deflected against the stationary housing vanes. The force of the fast-moving water striking the stationary housing causes the housing to try to rotate. A force-measuring load cell, shown in Figure 12-9, is used to keep the housing from rotating and to measure the torque the water exerts on the housing. The torque being measured at any RPM is controlled by changing the amount of water in the absorption unit. A water valve controls the flow of water through the absorption unit and, as the water flow increases, the amount of water in the absorption unit also increases.

There are two general types of dynamometers. One is an *engine dynamometer* which tests the engine by itself. Engine dynamometers are used to make engineering and performance studies on the engine, such as measuring and rating its horsepower. An engine under test on a dynamometer is shown in Figure 12-10.

The other type of dynamometer is called a *chassis dynamometer*. This unit is used in many service facilities for tuneup and diagnostic work. Motorcycles often are tested on a chassis dynamometer.

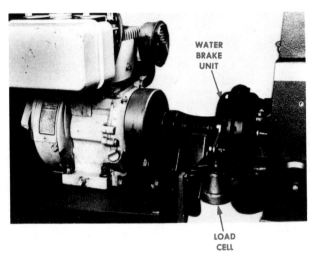

Figure 12-9. A force-measuring load cell.

Figure 12-10. A small engine being tested on a dynamometer.

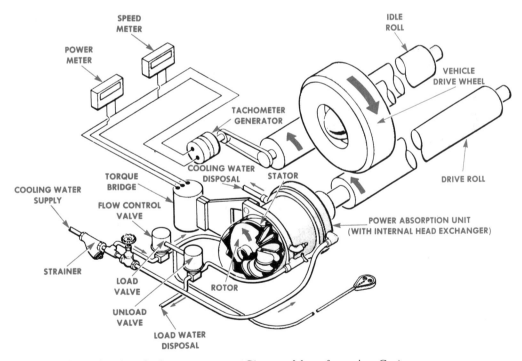

Figure 12-11. Operation of a chassis dynamometer. (Clayton Manufacturing Co.)

A chassis dynamometer allows the technician to duplicate road conditions and determine the vehicle's ability to perform on the road. It is useful for testing a number of motorcycle systems such as brakes, drive lines, and fuel systems as well as power output and engine condition. The chassis dynamometer puts the motorcycle to work exactly as it works on the road and continuously measures its ability to work.

The dynamometer is able to simulate road conditions by placing the rear wheel of the motorcycle between two large rollers. The shaft of the forward or drive roller is attached to a power absorption unit as shown in Figure 12-11. The power absorption unit works like a hydraulic fluid coupling. It has a rotor with blades which throws fluid forward. A stator with stationary blades receives the force of the fluid flow from the rotor. The power absorption unit acts like a brake. The more fluid in the unit, the more power required to revolve the rotor at any given speed. With a given amount of fluid, the faster the rotor revolves, the more power is required.

The motorcycle engine is started and the rider puts it in gear. The rear wheel drives the rollers. The operator, by maintaining a constant amount of fluid in the power absorption unit, may adjust any desired load on the rear wheels. The load may be changed by varying the amount of fluid within the system. The control is accomplished by two valves operated remotely from a portable control unit in the hand of the operator. By pushing an "on" button, the technician allows fluid to flow into the power absorption unit to simulate any type of road required from a level highway through moderate hills to the steepest grades. When the technician pushes the "off" button, fluid flows out of the power absorption unit, decreasing the load or steepness of the hill.

Torque and Horsepower Curves

When torque is measured on a dynamometer, it is recorded on a graph. Horsepower, calculated mathematically using torque and RPM also is charted on a graph. The graph shown in Figure 12-12 is typical. Horsepower figures are shown along one side of the graph and torque in pound-feet along the other side. The bottom of the graph has engine speed in revolutions per minute. This kind of graph allows both torque and horsepower to be compared to engine operating speed.

Horsepower and torque curves result when horsepower and torque are measured or calculated on a dynamometer. When an operator observes a certain amount of torque on the dynamometer instruments at a particular RPM, the operator puts a mark on the graph that corresponds to the torque and RPM. When the test is complete, all the marks are connected together and curves such as those shown in Figure 12-12 result.

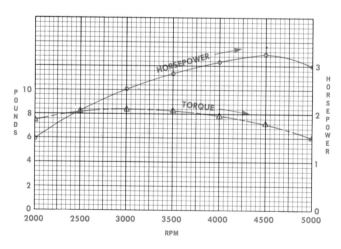

Figure 12-12. Horsepower and torque curve. (Go-Power Corp.)

The horsepower curve in Figure 12-12 is common to most engines. The horsepower does not start at zero because an engine will not run at zero speed. Therefore, the curve is cut off at the bottom. Horsepower increases as the engine speed and load increase. The graph shows that this engine reaches its maximum horsepower of 3.3 at 4,000 RPM. An engine is capable of running faster than the speed at which it reaches maximum horsepower. The horsepower begins to decrease after reaching the maximum point. The reasons for the decrease are related to the torque curve.

The torque curve shows the load-carrying ability of the engine at different speeds in pound-feet. The relationship between the torque curve and the horsepower curve shows how the engine will perform at different loads and speeds.

As shown in Figure 12-12, the horsepower curve continues to climb as the engine speed increases until maximum horsepower is reached. This is also true with the torque curve, but the torque curve will reach its maximum point much earlier. Notice in Figure 12-12 how the torque curve drops after it reaches its peak point (maximum) at 3,000 RPM.

The torque of which most engines are capable will change widely over the normal range of crankshaft speeds. At very low speeds—200-300 RPM — an engine develops only enough torque to keep itself running without any extra load. The next torque — the reserve beyond what is needed to keep the engine running — is practically zero.

As engine speed and load increase, torque will increase until it reaches the speed where torque will peak. This is where the manufacturer rates the torque. It will be very near the most efficient operating speed of the engine. At this point the cylinders are taking in the biggest and most efficient air-fuel mixture, and the exhaust gases in the cylinder are being forced out most effectively.

The torque curve drops off rapidly after its peak. At higher engine RPM there is less time for the air-fuel charge to enter the cylinder and less time for the exhaust gases to leave the cylinder. This results in a weaker push on the pistons and less torque. Other factors contribute to the drop in torque, such as internal engine friction and effective loss of the energy used in pumping in air

and fuel and exhausting it. These losses are described as pumping losses.

The horsepower curve is directly affected by the torque curve. This is due to the fact that torque is one of the elements in the horsepower formula:

$$\text{Horsepower} = \frac{\text{Torque} \times \text{RPM}}{5252}.$$ The

reason that the horsepower curve does not directly correspond with the torque curve is that it also is affected by another element, time. Power, remember, is the speed or rate at which work is done. The horsepower curve is able to increase past the peak of the torque curve because the engine RPM increases beyond this point. Eventually, however, the torque drops off so much that even more RPM cannot hold the horsepower curve up.

Horsepower Ratings

Even though there is general agreement about what horsepower is, there are some differences of opinion about how it is to be rated or specified. When horsepower is measured at the flywheel of an engine, it is described as brake horsepower, abbreviated BHP. The term *brake* comes from a device invented in 1821 by a man named Prony. It was called a Prony friction brake dynamometer. The Prony brake was a device that was wrapped around an engine flywheel to absorb energy and to measure the amount of energy absorbed.

The brake horsepower of an engine can be measured on a dynamometer with all of the engine's accessories driven from an external power source. This results in the maximum obtainable horsepower. Horsepower measured this way is called *gross horsepower*.

On the other hand, the horsepower can be measured with the engine driving all the necessary accessories, such as the standard exhaust and air cleaner, alternator, water pump, fan and oil pump. Figures obtained in this way are called the *net rating* and naturally are lower than the gross figures.

The horsepower of an engine is affected by such things as barometric pressure and air temperature. Horsepower readings taken directly from dynamometer readings are called *observed horse-power*. Observed horsepower is corrected to standard atmospheric conditions. This is called *corrected horsepower*. The correction factors may be those set up by SAE (The Society of Automobile Engineers) for the customary system or DIN, for the metric system. Horsepower corrected in this manner is called SAE horsepower or *DIN horsepower*. Since the metric standard is smaller than the customary horsepower unit, DIN ratings are slightly higher for a given engine than SAE ratings.

Indicated horsepower is an engineering measurement seldom used outside of a factory or laboratory. It is a measurement of the power delivered by the expanding gas to the piston inside the cylinder. It does not take into consideration the friction losses within the engine but is the total power developed in the cylinders. The usual formula is:

$$IHP = \frac{P \, L \, A \, N \, K}{33,000}$$

where P = means effective pressure in the cylinder in pounds per square inch.

   L = length of stroke in feet
   A = area of one cylinder
   N = number of cylinders
   K = a constant × RPM

(For a two-stroke, K = 1 × RPM, while for a four-stroke, K = 2 × RPM.)

The customary system of ratings known as taxable horsepower, which has been abandoned, bore almost no relationship to actual output. Taxable horsepower was based upon a calculation involving engine bore and number of cylinders. It was used for tax purposes.

Horsepower also may be measured with an engine installed in the vehicle. Horsepower measured at the rear wheels of a vehicle is described as *road* or *chassis horsepower*. This figure, of course, will be much lower than the engine ratings.

ENERGY

Energy in any form, is the ability to do work. The ability to do work is, of course, necessary to develop horsepower. The different kinds of energy are chemical, thermal and mechanical.

Mechanical energy is measured by the work a body can do. A raised weight has energy that is stored up for later use. If the weight is dropped, it can be made to do work. For example, the weight could be made to raise another weight, compress a spring or pull a rope or cable.

In an engine, the energy used to develop horsepower comes from the fuel. A fuel such as gasoline has a considerable amount of energy. The energy in gasoline is stored in the form of chemical energy. When the gasoline is burned with oxygen in the engine, the chemical energy is released as thermal energy.

Efficiency

The purpose of any engine or machine is to convert energy into useful work. How well the machine does at converting energy into work is a measure of its efficiency. Efficiency is the ratio of energy supplied to the work produced. The question is, how much work does a machine deliver for the amount of energy put into it? The basic equation for efficiency is:

$$\text{Efficiency} = \frac{\text{work output}}{\text{energy in}} \times 100\%$$

No machine or device is 100 percent efficient because of lost energy through heat and friction. In an internal combustion engine, only about 25 percent of the total energy of the fuel is converted into useful work.

Thermal Efficiency. The internal combustion engine converts the chemical energy of gasoline into thermal energy. It is important to know how efficiently this is accomplished. Thermal efficiency is a measure of the percentage of heat energy available in the fuel that actually is converted into power at the crankshaft.

Volumetric Efficiency. Another important type of efficiency rating in internal combustion engines is called *volumetric efficiency*. Volumetric efficiency is the relationship of the actual volume of a cylinder to the volume which is filled during engine operation. At high engine speeds, the valves are open for such a short time that the cylinders are not completely filled. This, plus friction temperature and inertia, causes the cylinders to be filled to less than their capacity. Volumetric efficiency may be calculated with an engine dynamometer and air flow equipment.

Brake Mean Effective Pressure. Another important measurement of engine efficiency is called *brake mean effective pressure,* abbreviated *BMEP.* BMEP is the average effective pressure exerted on the piston during one operating cycle. Pressure used in drawing fuel into the cylinder and compressing it, pressure required to exhaust the burnt exhaust gases, and internal friction are subtracted from the power delivered during the power stroke. The result is BMEP. BMEP may be calculated on a dynamometer.

NEW TERMS

bore: The diameter of the cylinder.

brake horsepower: Horsepower measured at the engine's flywheel. Abbreviated BHP.

brake mean effective pressure: The average pressure exerted on the piston during one operating cycle. Abbreviated *BMEP.*

compression ratio: The amount the air-fuel mixture is compressed during the compression stroke, compared to its original volume.

displacement: The volume swept or displaced by the pistons of an engine.

dynamometer: Equipment used to measure torque and calculate horsepower.

efficiency: How well a machine such as an engine converts energy into useful work.

energy: The ability to do work.

force: A push or a pull.

horsepower: Term used to describe the power developed by an engine. One horsepower is equal to 33,000 foot-pounds of work per minute.

horsepower ratings: Different ways in which horsepower is measured and specified.

indicated horsepower: A laboratory horsepower measurement based upon the power developed in the engine's cylinders.

linear horsepower: Horsepower used to pull in a straight line.

power: The speed at which work is done.

rotary horsepower: Horsepower developed in a rotary motion such as by an engine's crankshaft.

stroke: The movement of the piston in the cylinder, controlled and measured by the offset of the crankshaft.

thermal efficiency: How well an engine changes the chemical energy in gasoline to heat energy.

torque: A turning or twisting effort or force.

volumetric efficiency: The ratio of an engine's cylinder volume to the volume filled by air and fuel during engine operation.

work: What is done when a force travels through a distance.

SELF CHECK

1. Define the term *bore.*
2. Define the term *stroke.*
3. Explain the term *displacement.*
4. Define *compression ratio.*
5. Define *force.*
6. Write the formula for work.
7. Define the term *power.*
8. Define a *horsepower.*
9. What is rotary horsepower?
10. Explain how horsepower is measured on a dynamometer.
11. What is a torque curve?
12. What is a horsepower curve?
13. What is efficiency?
14. Define *thermal efficiency.*
15. Define *volumetric efficiency.*

DISCUSSION TOPICS AND ACTIVITIES

1. Look up the horsepower and torque specifications for a small engine you own. At what RPM is horsepower the highest? Where is torque highest?
2. What limits the horsepower and torque of an engine? How could they be increased?

A motorcycle has a small engine.
(U.S. Suzuki Motor Corp.)

part 5
troubleshooting, maintenance and tune-up

Now that we have studied how small engines work, we are ready to learn how they are repaired. The first area we will study is troubleshooting. Troubleshooting involves following a step-by-step procedure to find out what is wrong with an engine. We will also learn how to do periodic maintenance on small engines. Periodic maintenance involves service jobs done at regular intervals to keep the engine running correctly.

Both the ignition and fuel systems require periodic service and adjustment in order to operate correctly. We call these service jobs a tune-up. You will learn how to do a tune-up on a small engine.

unit 13

troubleshooting small-engine problems

Small engines are designed and built to operate reliably for long periods of time. After a while the engine may become difficult to start. It may begin to run roughly or to make noise. It may fail to start at all. The first step in repairing an engine is to find out what parts are causing the problem. Following a step-by-step procedure to find the source of the problem is called *troubleshooting*. A troubleshooting procedure is based on a systematic step-by-step check of the most likely problems. Unless a logical procedure is followed, a great deal of time can be lost looking in the wrong places. In this unit we will see how to troubleshoot a small engine.

LET'S FIND OUT: **When you finish reading and studying this unit, you should be able to:**
1. **Define the term *troubleshooting*.**
2. **Describe how to troubleshoot small-engine fuel problems.**
3. **Describe how to troubleshoot small-engine ignition problems.**
4. **Describe how to troubleshoot small-engine compression problems.**
5. **Explain how to read a troubleshooting chart.**

THREE BASIC PROBLEMS

In order to start and run, an engine must have fuel, ignition and compression. If the engine does not get enough fuel, or if it gets too much, it will not start. A loss of ignition or ignition that occurs at the wrong time will not allow the fuel mixture to ignite, and the engine will not start. Unless the air-fuel mixture is compressed tightly on the compression stroke, there will be no explosion for a power stroke.

When an engine will not start, the mechanic

must always look for the three basics. If the engine has fuel, ignition and compression, it will run. In the next section we will see how to trouble-shoot for fuel, ignition and compression.

IGNITION SYSTEM TROUBLESHOOTING

An engine that will not start or run may have an ignition system problem. An engine with a magneto problem will not produce the high-voltage spark. Without a spark the engine will not start or run. You know how the magneto works. You can find out if the magneto is working. BE CARE-FUL. *Remember that the spark plug wire carries high voltage. Do not touch the end that connects to the spark plug. If you touch this end when turning the flywheel, you could get on electric shock.*

To check for spark, take the spark plug wire off the spark plug. Most spark plug wires have an end that fits around the spark plug terminal. To remove it, just pull th wire straight off the terminal.

A spark tester, like that shown in Figure 13-1, may be used to look for ignition. Hook the spark plug wire to the spark tester. Hold the spark tester by the handle. Touch the other end of the tester to the cylinder head, as shown in Figure 13-2. Use your other hand to spin the flywheel. A spark should jump the gap of the tester. A spark jump-

ing the tester gap is shown in Figure 13-3. If a spark jumps the tester gap, the magneto is working. If the engine will not start, the problem is in some other part. We will show you how to check these other parts later.

(*Note:* Some CD (capacitive discharge or solid state) ignition systems can be damaged by testing this way. Check with the manufacturer before testing a CD system for spark.)

It is possible to check for spark without a spark tester. Remove the spark plug wire from the spark plug. Hold on to the wire away from the terminal end. *Do not touch the terminal or you will be shocked.* Hold the end of the spark plug wire a short distance away from the spark plug shell, as

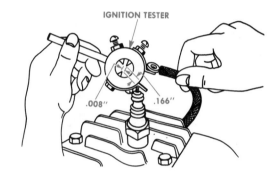

Figure 13-2. Using a spark tester to check for spark. (Briggs & Stratton Corp.)

Figure 13-1. A spark tester is used to check for ignition.

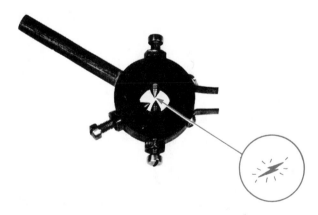

Figure 13-3. A spark should jump across the tester if the magneto is working.

Figure 13-4. Checking for spark without a tester.

Figure 13-5. Checking for ignition on a motorcycle or moped.

shown in Figure 13-4. Spin the flywheel. A bright, hot spark should jump from the spark plug wire to the spark plug. If you see the spark, the magneto is working. Hook up the spark plug wire again.

A motorcycle or moped is tested for ignition in exactly the same way. Disconnect the ignition wire from the spark plug and position it a short distance from the cylinder, Figure 13-5. The engine is then cranked over with the ignition key switch turned on. The spark should have a blue-white color. If there is no spark, or if it is faint or yellow in color, the ignition system is not working correctly.

Outboard engines are tested for ignition in the same way. A large cover must be removed from the top of the power head. The cover often is held in place with several screws or quick-release clamps. With the cover removed, the spark plugs and spark plug wires are accessible, Figure 13-6. Check the engine for spark just like any other engine.

If an engine does not have a good strong spark, the ignition system will require further testing and service. The mechanic must test and repair as necessary each part of the ignition system. We will present the procedures for servicing the ignition system in the unit on tuneup.

FUEL SYSTEM TROUBLESHOOTING

An engine must have fuel to start and run. It must have just the right amount of fuel to run smoothly. A fuel problem may make the engine hard to start. It may not even start at all. Too much or too little fuel will cause the engine to run roughly. *You will be working around gasoline. Gasoline can explode! Make sure there are no open fires or flames near where you are working. Do not work in a closed area. Make sure the shop area is well ventilated.*

The first step in finding a fuel problem is to take out the spark plug. When the spark plug is out, look at the area around the electrodes. This area will give us a clue about how the fuel system is working.

Does the spark plug look like the one in Figure 13-7? The fuel mixture for this engine is just right. The spark plug is brown or tan in color. If the spark plug does not look like this, there may be a fuel problem.

Figure 13-6. The power head cover is removed to get to the outboard ignition system.

Figure 13-7. The right air-fuel mixture leaves the spark plug brown or tan in color. (Champion Spark Plug Co.)

A fuel problem can let too much fuel get into the cylinder. An air-fuel mixture with too much fuel is hard to burn. The engine will run roughly. Fuel will soon cover the spark plug. A spark plug wet with fuel will not spark. The engine will not start. Does the spark plug look like the one in Figure 13-8? This engine is getting too much fuel. The spark plug is wet with gasoline. It has a black color.

One common cause of too much fuel is over-choking. Has the engine been running with the choke valve closed? Look at the choke valve. Is it open or closed? Move the choke lever, Figure 13-9, back and forth. Make sure the choke valve is not sticking.

The carburetor high-speed or low-speed adjustment screws may be out of adjustment. This can cause too much fuel to enter the cylinder. We will see how to adjust these screws later.

An engine that will not start or run may not have enough fuel. Does the spark plug look like the one in Figure 13-10? This spark plug is from an engine that is not getting enough fuel. The spark plug is dry. It is white in color.

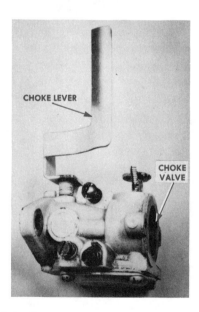

Figure 13-8. A black, wet spark plug means too much fuel. (Champion Spark Plug Co.)

Figure 13-10. A dry, white spark plug means not enough fuel. (Champion Spark Plug Co.)

Figure 13-9. Check the choke by moving the lever back and forth.

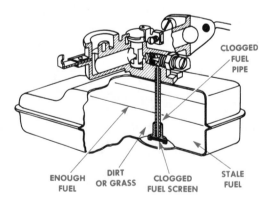

Figure 13-11. Look for these problems at the fuel tank. (Briggs & Stratton Corp.)

Remove the cap from the fuel tank and look inside. Is there fuel in the tank? How does the fuel smell? Gasoline that sits around too long gets stale. Stale gasoline smells like varnish and can plug up the carburetor.

Make sure the fuel tank valve, Figure 13-11, is open. If it is closed, fuel cannot go to the carburetor. Sometimes dirt, water or grass gets into the fuel tank and clogs the fuel line at the bottom of the tank. Some fuel tanks have a screen in the fuel line. The screen is used to keep dirt from going to the carburetor. This screen can be full of dirt so the fuel can't get through. A plugged fuel cap vent can stop fuel from going to the carburetor.

The carburetor high- and low-speed adjustment screws may be out of adjustment, not letting enough fuel into the cylinder. We will see how to adjust those screws in another unit.

Most outboard engines use a separate fuel tank, Figure 13-12. The fuel tank is nonpressurized. Fuel is pumped from the tank to the carburetor by the fuel pump. Priming is achieved by squeezing the primer bulb (part of the fuel line) several times or until pressure required to squeeze the bulb increases. The connector nearest the primer bulb must be connected to the fuel tank.

The tank air inlet and fuel outlet are seated until the supply line connector is plugged into the tank. When the fuel line is attached, two valve

Figure 13-12. An outboard fuel tank.

plungers are depressed, forcing the valves off their "O" ring seats.

The outboard tank should be checked to make sure there is enough fresh, properly mixed fuel. A pickup screen on top of the tank may be disassembled and inspected for blockage. The fuel should be checked to see that it contains no dirt, varnish or water.

Motorcycle and moped fuel systems are checked just like any other small engine. The spark plug is removed to determine if there is too much or too little fuel. The fuel tank, Figure 13-13, is located above the engine. The carburetor

Figure 13-13. Motorcycle fuel system. (U.S. Suzuki Motor Corp.)

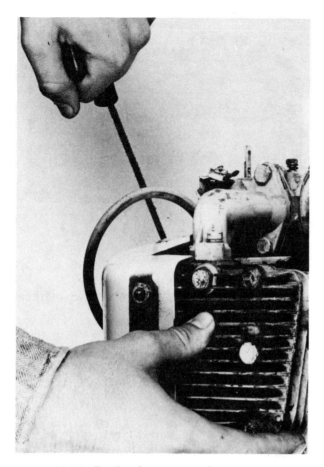

Figure 13-14. Feeling for compression.

Figure 13-15. Compression testing gage.

gets fuel from a gravity flow. If the fuel level is high enough, it will be visible in the tank after the cap is removed. If the level is very low, you can shake the motorcycle from side to side and listen for the sound of fuel in the tank. The fuel line to the carburetor may be removed to check for a clear fuel flow. A tank valve is used to shut off fuel flow from the tank when the engine is stopped. This valve must be open when the engine is running. A screen in the tank valve assembly can be checked for blockage.

COMPRESSION TROUBLESHOOTING

An engine may have ignition and fuel, but without enough compression it still will not run. The air-fuel mixture must be compressed tightly or there will be no power stroke. After the mechanic has made sure the engine has ignition and fuel, the next step is to check for compression.

The mechanic can make a simple test for compression. First the spark plug is removed. The spark plug wire is grounded properly away from the spark plug hole. The mechanic covers the spark plug hole with a thumb as shown in Figure 13-14. The engine is then cranked as if it were to be started. The mechanic should feel a definite, strong push against the thumb on each compression stroke. If not, the compression is too low to allow the engine to start.

The exact compression of the engine can be determined with a compression pressure gage, Figure 13-15. Compression pressure tests tell the condition of piston rings, valves, and head gaskets. By comparing results of the tests with manufacturer's specifications, the mechanic can tell if a cylinder is working correctly.

A compression gage is put in the spark plug hole of a cylinder, Figure 13-16. The compression pressure is then read while the engine is cranked. If the engine has more than one cylinder, this procedure is repeated for each cylinder. Reading should be taken with the engine at normal operating temperature to show the best ring and valve sealing under normal operating conditions. The carburetor throttle valve should be opened fully to allow atmospheric pressure to force a full mix-

Figure 13-16. Measuring the compression of a small engine.

ture into the cylinder. All spark plugs should be removed from a multiple-cylinder engine to prevent too much engine drag during cranking. All cylinders must be tested the same number of strokes for a correct comparison. Readings should be noted on the first compression stroke as well as the last stroke to fully determine engine condition.

A cylinder that is below specifications and varies more than 20 pounds per square inch or 1 kilogram per square centimeter from the highest cylinder reading is considered abnormal. Most four-stroke engines have a normal compression of around 75 PSI while many two-stroke engines have a normal compression of about 90 PSI. An abnormal reading, along with low-speed missing, indicates an improperly seated valve or worn or broken piston rings. Worn piston rings are indicated by low compression on the first stroke which tends to build up on the following strokes. A further indication of worn rings is an improved reading when several squirts of oil are added to the cylinder. Valve problems are indicated by a low compression reading on the first stroke which

does not rapidly build up on following strokes and is not changed by the addition of oil. Leaking head gaskets give nearly the same test results as valve problems.

TROUBLESHOOTING ENGINE NOISES

Another sign of engine trouble is an abnormal sound, especially a knock. Knocking noises usually mean the engine must be disassembled for major service. The experienced mechanic becomes expert at finding the causes of knocks by their sound and location. Engine accessories should be disconnected to make sure that the noise is inside the engine. If the engine has more than one cylinder, the mechanic can take the load off one cylinder at a time by disconnecting the spark plug wire. The loudness of the knocking will change when the spark plug in the cylinder with the problem is disconnected. Knocking can also result from too much or too little clearance between any moving surfaces in the engine — main rod bearings, piston skirts and piston pins. Knocks may also be caused by broken or fractured piston skirts and piston rings.

The experienced mechanic also can tell the trouble by the knock speed. A knock that occurs at half engine speed could be caused by a valve tappet or other valve train component.

Main bearing knocking, indicating too large a bearing clearance, usually occurs as a heavy thump and only when the engine is under load. A lighter knock when the engine is not under load may mean a bad connecting rod bearing.

FOLLOWING A TROUBLESHOOTING CHART

Many manufacturers of small engines provide a troubleshooting chart in their service manual. The troubleshooting chart is a guide the mechanic can follow in locating a problem. The chart usually lists the problems starting with the ones that occur most frequently. The possible causes are listed under each problem. The things the mechanic must do to repair the problem are listed next to each cause. When using a troubleshooting chart, the mechanic should follow each of the recommended steps. It sometimes helps to check off each step as it is finished. A troubleshooting chart is provided in this unit for most of the common small engines. It is always best, however, to find and use the chart designed for the exact engine on which you are working.

TROUBLESHOOTING CHART FOR SMALL TWO- AND FOUR-STROKE ENGINES

Cause	Remedy
ENGINE FAILS TO START OR STARTS WITH DIFFICULTY	
On-Off switch Off	Turn switch to On.
No fuel in tank	Fill tank with clean, fresh fuel.
Shut-off valve closed	Open valve.
Obstructed fuel line	Clean fuel screen and line. If necessary, remove and clean carburetor.
Tank cap vent obstructed	Open vent in fuel tank cap.
Water in fuel	Drain tank. Clean carburetor and fuel lines. Dry spark plug points. Fill tank with clean, fresh fuel.
Engine over-choked	Close fuel shut-off and pull starter until engine starts. Reopen fuel shut-off for normal fuel flow.
Improper carburetor adjustment	Adjust carburetor.
Loose or defective magneto wiring	Check magneto wiring for shorts or grounds; repair if necessary.
Faulty magneto	Check timing, point gap, and, if necessary, overhaul magneto.
Spark plug fouled	Clean and regap spark plug.
Spark plug porcelain cracked	Replace spark plug.
Poor compression	Overhaul engine.
ENGINE NOISES	
Carbon in combustion chamber	Remove cylinder head or cylinder and clean carbon from head and piston.
Loose or worn connecting rod	Replace connecting rod
Loose flywheel	Check flywheel key and keyway; replace parts if necessary. Tighten flywheel nut to proper torque.
Worn cylinder	Replace cylinder.
Improper magneto timing	Time magneto.
ENGINE MISSES UNDER LOAD	
Spark plug fouled	Clean and regap spark plug.
Spark plug porcelain cracked	Replace spark plug.
Improper spark plug gap	Regap spark plug.
Pitted magneto breaker points	Clean and dress breaker points. Replace badly pitted breaker points.
Magneto breaker arm sluggish	Clean and lubricate breaker point arm.
Faulty condenser (except on Tecumseh Magneto)	Check condenser on a tester replace if defective.
Improper carburetor adjustment	Adjust carburetor.

Cause	Remedy
Improper valve clearance (four-cycle engines)	Adjust valve clearance.
Weak valve spring (four-cycle engine)	Replace valve spring.
Reed fouled or sluggish (two-cycle engine)	Clean or replace reed.
Crankcase seals leak (two-cycle engine)	Replace worn crankcase seals. Some two-cycle engines have no lower seal. Check bearing surface of bottom half of crankcase.

ENGINE LACKS POWER

Cause	Remedy
Choke partially closed	Open choke.
Improper carburetor adjustment	Adjust carburetor.
Magneto improperly timed	Time magneto.
Worn piston or rings	Replace piston or rings.
Lack of lubrication (four-cycle engine)	Fill crankcase to the proper level.
Air cleaner fouled	Clean air cleaner.
Valves leaking (four-cycle engine)	Grind valves.
Reed fouled or sluggish (two-cycle engine)	Clean or replace reed.
Improper amount of oil in fuel mixture (two-cycle engine)	Drain tank; fill with correct mixture.
Crankcase seals leaking (two-cycle engine)	Replace worn crankcase seals. Some two-cycle engines have no lower seal. Check bearing surface of crankshaft.

ENGINE OVERHEATS

Cause	Remedy
Engine improperly timed	Time engine.
Carburetor improperly adjusted	Adjust carburetor.
Air flow obstructed	Remove any obstructions from air passages in shrouds.
Cooling fins clogged	Clean cooling fins.
Excessive load on engine	Check operation of associated equipment. Reduce excessive load.
Carbon in combustion chamber	Remove cylinder head or cylinder and clean carbon from head and piston.
Lack of lubrication (four-cycle engine)	Fill crankcase to proper level.
Improper amount of oil in fuel mixture (two-cycle engine)	Drain tank; fill with correct mixture.

ENGINE SURGES OR RUNS UNEVENLY

Cause	Remedy
Fuel tank cap vent hole clogged	Open vent hole.
Governor parts sticking or binding	Clean and if necessary repair governor parts.
Carburetor throttle linkage or throttle shaft and/or butterfly binding or sticking	Clean, lubricate, or adjust linkage and deburr throttle shaft or butterfly.

ENGINE VIBRATES EXCESSIVELY

Cause	Remedy
Engine not securely mounted	Tighten loose mounting bolts.
Bent crankshaft	Replace crankshaft.
Associated equipment out of balance	Check associated equipment.

TROUBLESHOOTING CHART FOR A TWO-STROKE MOPED OR MOTORCYCLE ENGINE

STAGE 1 — ENGINE WON'T RUN

Fuel

Remove plug: Is it dry? And/or
Remove gasoline line: Does the fuel flow out?

Compression

Preparation: Remove spark plug.
What to do: Put a finger tightly over the hole, and kick over the engine. Does it blow your finger off the hole?

Ignition

Preparation: Remove spark plug and lay it on cylinder fins with high-tension lead attached.
What to do: Kick over the engine, watching the electrodes. Is there spark?

STAGE 2 — ROUGH, OCCASIONAL POPS (COMMON TOOLS)

By now, we hope to have discovered some sort of deficiency in one of the three areas. Proceed by checking further into this malfunction hint while being alert for various symptoms which may pinpoint a deficiency in this area or in one of the others as well. Do not be afraid at any time to jump over into another area and check again, starting from the simple stages and progressing.

Fuel

Rich — If the plug is wet, clean it off and clear out the engine as much as possible by turning it over. Subsequently, check for restriction of intake air and/or excessive fuel supply as indicated by excessive wetness around the carburetor.

Lean — If the plug is dry, replace and attempt by choking or various means to flood the engine or at least create a wet appearance on the spark plug.

Compression

Repeat the above process, holding your finger as tightly as possible over the spark plug hole to see if it makes a high-pressure squeaking noise. Remove your finger and rotate the engine both slowly and rapidly, and listen for strange noises. If possible, look into the hole for unusual colorations or deposits.

Ignition

Hold the high-tension lead in your hand and kick over the engine so that the spark plug must jump about ¼-inch or so to the head. Is this spark fat and juicy?

STAGE 3 — INTERMITTENT (REQUIRING SPECIAL TOOLS)

You are still dealing with an engine that does not run at all or just barely pops, but not enough to keep going. By now you should have a pretty fair indication of which one of the areas to pursue further. Meanwhile, stay alert for other clues such as increasing or decreasing wetness, unusual pops, breaking, banging or hissing sounds. Also watch for irregularities in the strength or frequency of the spark, or any other clue which will help identify the problem area.

Fuel

Rich — Check to see if carburetor is flooded internally. Gasoline should not spill over from the carburetor bowl. Also check to make sure that the choke is not in operating position. Also check the air filter and air ducting system to make sure that they are not excessively restricted.

Lean — See that the carburetor is tightly mounted and has no loose parts such as the slide cover or bowl. While kicking the engine over, listen carefully for any bubbling or hissing noise which would indicate a potential air leak, either in the crankcase or at the cylinder base.

Compression

Using a compression gage, check to see if the engine has the required 115-135 pounds pressure. Remove the head and check for even, dark discoloration inside the head and atop the piston. Especially check the condition of the cylinder wall with the piston at BDC (before dead center).

Ignition

1. Make an instrument spark test to determine if secondary coil output is the required minimum of 7mm.
2. Check the contact point surfaces, looking carefully for excessive pitting, gray or oxidized surface, and also for free and easy movement of points. Also turn the engine over while watching the points to be sure that they actually do make and break contact.
3. Give the timing a rough check to make sure the spark plug actually is trying to fire slightly ahead of top dead center.

STAGE 4 — POOR POWER (MORE SPECIAL TOOLS)

Fuel

Rich —
Disassemble the carburetor to make sure that the main jet is of approximate proper size and is in place. Check the float by shaking it to see if there is fuel inside which would cause it to sink. Also check the float level to make sure that it is the proper measurement.

Lean —
Check to make sure that the jet needle has not lost its clip and dropped to the bottom. Also check for adequate flow of fuel through the float needle valve when opened. Visually inspect all jets and passages for signs of restriction or the presence of dirt or corrosion.

General:
1. If the machine has two carburetors, check for full and synchronized throttle action.
2. If the machine has a rotary valve, remove the cover if possible and check for proper sealing on both sides of the valve. Also check for approximate opening of the crankcase intake port at slightly more than ¼ of a turn BTDC (before top dead center), closing slightly after TDC (top dead center).

Compression

Remove the cylinder and check for the following:
1. Scoring, scratches or deposits in the cylinder.
2. Piston seizure or excessively loose fit of pistons in cylinder.
3. Rings stuck in their grooves in any way, or scored, chipped or having excessive end gap.

Ignition

1. With the ignition coil hooked up to a bench testing apparatus, allow it to operate for a half-hour or so to get thoroughly warm inside to see if there is any breakdown of the spark after continuous operation.
2. Test the condenser in the manner approved by instructions.
3. If the machine has a direct current primary system, make a voltage check with a voltmeter of all circuits of the primary ignition system.
4. Point movement. Double check point movement to make sure that the movable point can flop freely and that it does have the specified spring pressure as measured with a gage.

STAGE 5 - FINE TUNING (TOOLS AND PARTS)

Fuel

Whether rich or lean, observe the following:
1. Spark plug reading. Using a new plug, make the required spark plug reading for each phase of carburetor opening and re-jet accordingly.
2. Fuel flow. Double check all gasoline and air passages and openings. If carburetion problems persist, boil out the carburetor in approved carburetor cleaning solutions and thoroughly blow through all passages with compressed air.
3. Vibration. Check all tanks, lines, carburetors and fittings for vibration which could cause frothing or other fuel stoppage.
4. Crankcase test. Perform this check carefully. Thoroughly check the position and condition of all crankcase seals, remembering that poor bearings or excessive vibration can cause these to leak even though their appearance is good.

Compression

1. Remove the rings and check the piston grooves for signs of carbon formation thick enough to cause restriction of ring movement. Clean as necessary.
2. Using precision measuring instruments, check cylinder taper or out-of-roundness, neither of which should be more than .002″ at the most. Also check for piston clearance in the cylinder, which should not be more than .002″ per inch of bore.

Ignition

1. Set the point gap to the specified dimensions at the widest opening of the point cam.
2. Set the timing accurately with precision instruments.
3. Replace the spark plug with a new one of either correct heat range or one which has been proven to be needed by actual spark plug reading.
4. Carefully test or replace the high-tension ignition wire.
5. Attach a strobe light to the machine and run the engine through the RPM range, carefully looking for deviations, fluctuations, or maverick conditions in the spark timing. Suspect any bad bearings or loose engine condition which could cause spark deviations.
6. Carefully test and/or replace every ignition component not mentioned above.

TROUBLESHOOTING CHART FOR AN OUTBOARD ENGINE

TROUBLE	POSSIBLE CAUSE
1. MOTOR WILL NOT START	A. FUEL SYSTEM Fuel line improperly connected Engine not primed Speed control not advanced (throttle closed) Engine flooded Old fuel Clogged fuel filter Choke not closing completely Choke spring broken or disconnected Fuel system faulty B. IGNITION SYSTEM Bad spark plugs Spark plug leads crossed or reversed Linkage improperly adjusted Sheared flywheel key Ignition system faulty C. NEUTRAL START Gear shift not in neutral
2. LOSS OF POWER — (Providing Ignition is OK)	A. POWERHEAD Carburetor and magneto not synchronized Throttle control lever bent (won't advance) Air leak at manifold gaskets — warped manifold (backfires) Broken leaf valves (backfires) Worn lower main bearing seal Excessive carbon on pistons and cylinder head Stuck piston rings, or scored cylinder or piston Recirculating system plugged B. CARBURETOR Poor fuel mix — too much lubricant Linkage screws loose Choke not operating Inlet needle and seat worn or sticky Incorrect carburetor float setting Boss gasket missing Dirt in high speed orifice Altitude horsepower loss C. FUEL PUMP AND TANK Faulty fuel hose (clamps or seals) (kinked) Fuel pump filter restricted Fuel tank filter plugged Fuel and vent valves not opening Diaphragm leaking or damaged Fuel system hoses plugged "O" ring damaged in fuel line connector

TROUBLE	POSSIBLE CAUSE

POSSIBLE CAUSE

D. EXHAUST GAS ENTERING CARBURETOR
Exhaust cover screws leaking
Cover plate gasket damaged
Damaged exhaust housing seal
Adapter gaskets leaking
Cracked exhaust housing
Exhaust tube to cylinder screws loose or missing

E. OVERHEATING POWERHEAD
Exhaust cover gasket leaking
Inner exhaust cover leaking
Powerhead gasket improperly installed or damaged
Head gasket leaking (warped head) (water in cylinders)
Water intakes obstructed
Pump housing air bleed restricted
Water passages obstructed
Pump housing air bleed restricted
Water passages obstructed
Pump plate not sealing (bottom)
Pump impeller damaged
Pump housing or plate worn
Pump housing seal worn (driveshaft grooved)
Water tube grommet damaged

F. LOWER UNIT
Propeller hub slipping
Bent or worn propeller
Bent gear housing or exhaust housing

G. EXHAUST GASES ENTERING COOLING SYSTEM
Pump impeller plate not sealing (bottom)
Damaged water tube grommets or "O" rings
Pump housing seal damaged (5" adapter seals)
Exhaust tube to plate gasket damaged

3. MOTOR MISFIRES (Providing fuel System and Carburetor are OK)

A. SPARK PLUGS
Cover or inner terminal damaged (spark plug point out of H.T. lead)
Broken wires
Loose — low torque
Incorrect heat range
Defective (cracked insulator)

B. IGNITION
Connectors improperly joined
Coil leads loose
Charge coil faulty or loose (rope model)
Sensor coil faulty or loose (rope model)
Sheared flywheel key
Throttle cam improperly adjusted
Loose armature plate screws
Power pack faulty

(Continued on next page.)

TROUBLE	POSSIBLE CAUSE

4. POOR PERFORMANCE ON BOAT

    A. MOTOR ADJUSTMENTS
       Incorrect propeller
       Incorrect tilt angle
       Remote controls incorrectly adjusted
    B. CAVITATION
       Protruding hull attachments
       Keel too long
       Bent propeller (vibration)
       Transom too high
    C. BOAT
       Improper load distribution
       Marine growth on bottom
       Added weight (water absorption)
       Hook in bottom
       Catamaran (single engine) — venturi effect

5. STARTER MOTOR WILL NOT OPERATE

    A. STARTING CIRCUIT
       20-amp fuse blown
       Loose or corroded battery connections
       Safety switch inoperative (loose)
       Throttle advanced too far
       Poor or broken battery connections
       Weak or shorted battery
       Open circuit in solenoid
       Defective key or starter switch
    B. STARTER ASSEMBLY
       Jammed starter drive
       Damaged starter drive parts
       Worn brushes
       Broken brush spring
       Burned commutator
       Broken field terminal
       Shorted or open windings — armature or field
    C. EXCESSIVE STARTER CURRENT DRAW
       Worn or dry armature shaft bearings
       Excessive friction in engine
       Brushes not seating
       Dirty or corroded commutator
       Loose pole pieces
       Shorted armature or field
       Bearing heads buckled

NEW TERMS

compression gage: A tester used to determine if an engine has enough compression.

knock: A metallic noise in an engine that generally indicates a problem.

troubleshooting: The logical step-by-step procedure used to locate an engine problem.

troubleshooting chart: A chart listing problems along with causes and cures.

SELF CHECK

1. What is a spark tester?
2. Why must you never touch the spark plug wire terminal?
3. Explain how to test for spark with a spark tester.
4. Explain how to test for spark without a spark tester.
5. What breaker point problems might stop the magneto from working?
6. What does a spark plug look like when the air-fuel mixture is just right?
7. How does a spark plug look when there is too much fuel?
8. How does a spark plug look when there is not enough fuel?
9. What problems should we look for at the fuel tank?
10. Explain how to check an engine for compression.

DISCUSSION TOPICS AND ACTIVITIES

1. Describe the steps you would follow to find a problem in a small engine you have.
2. Have a friend "bug" a small engine. See how long it takes you to find the problem.

unit 14
periodic maintenance

In order to work correctly for a long period of time, a small engine requires periodic maintenance. Periodic maintenance involves service jobs done at regular intervals. The intervals usually are specified in months or hours of operation. The necessary jobs and the required intervals are provided in an owner's manual supplied by the engine manufacturer. In this unit we will study how to do the most common periodic maintenance jobs.

LET'S FIND OUT: **When you finish reading and studying this unit, you should be able to:**
1. **Describe how to check engine oil.**
2. **Describe how to change engine oil.**
3. **Explain how to service an oil cooling system.**
4. **List the steps to follow in servicing an air cleaner.**
5. **Explain how to prepare an engine for storage.**

CHECKING ENGINE OIL

The engine should have just the right amount of oil. The oil should come up far enough in the crankcase for the dipper or pump to pick it up. The height of the oil in the crankcase is called the *oil level*. The oil level should be checked each time the engine is used.

To check the oil level, find the oil filler plug. There are usually two plugs on the outside of the engine. One is used to drain the oil. It is on the very bottom of the crankcase, as shown in Figure 14-1. The higher plug is the oil filler plug. The oil filler plug usually is made of plastic. Often it is shaped like a wing nut. It can be removed by hand.

Put the engine on a level bench. Remove the oil filler plug by turning it to the left. Look into the oil filler plug hole. If the oil is not at the recommended level, add oil to the crankcase until it comes to the top of the oil filler plug hole. The recommended type of oil to use in an engine may be found in the manufacturer's service manual. Always make sure you use the correct oil in the engine. Replace the oil filler plug by turning it to the right. Be careful not to damage the threads.

OIL
FILLER
PLUG

OIL
DRAIN PLUG

Figure 14-1. An engine has a filler and a drain plug. (Briggs & Stratton Corp.)

OIL
DRAIN
PLUG

Figure 14-2. The drain plug is on the bottom of the crankcase. (Briggs & Stratton Corp.)

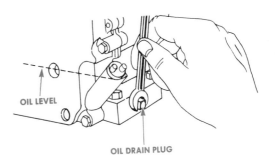

OIL LEVEL

OIL DRAIN PLUG

Figure 14-3. The drain plug is removed to drain the old oil. (Briggs & Stratton Corp.)

CHANGING ENGINE OIL

As the engine runs, oil in the crankcase gets dirty. Oil with dirt in it circulated around the engine parts will quickly scratch and damage the engine parts. The dirty oil must be drained out of the engine at intervals recommended by the manufacturer. First find the oil drain plug. The drain plug is on the bottom of the crankcase, as shown in Figure 14-2.

Remove the plug over a pan to catch the old oil. Use a wrench to remove the drain plug. Turn the plug to the left as shown in Figure 14-3. As soon as the drain plug is removed, oil will pour out. Tip the engine so that all the oil drains out. Replace the drain plug. Use the wrench to tighten it. Turn it to the right or clockwise.

Put the engine on a level bench. Remove the oil filler plug. Place a funnel in the oil filler plug hole. Pour oil into the crankcase through the funnel, as shown in Figure 14-4. Fill the crankcase up to the top of the oil filler plug hole. Replace the oil filler plug. Wipe up any oil you have spilled.

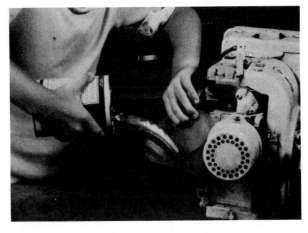

Figure 14-4. Fill the crankcase with oil through the filler plug hole.

CLEANING COOLING FINS

Many engines run where there is a lot of dirt or grass in the air. A power lawn mower and a chainsaw both have this problem. Dirt or grass in the air around the engine is pulled into the blower housing. The grass and dirt stick to the cooling fins. The fins can become so plugged up that air cannot go through. Heat will not be taken away, and the engine parts will run too hot. If we do not clean the fins, the engine parts can be damaged. *Engine parts are hot when the engine has been running. Let the engine cool down before you touch it.*

First remove the blower housing. The blower housing usually is held to the engine with three bolts or screws. One is at the top of the blower housing. The other two are along the side, as shown in Figure 14-5. Find the bolts. Use your wrench to take the bolts out, Figure 14-6, by turning the bolts to the left.

Lift off the blower housing as shown in Figure 14-7. Wipe out the blower housing with a rag. This will shake grass and dirt out of the housing.

Look at the cooling fins on the cylinder head and cylinder. Are they plugged up with grass or

Figure 14-6. Remove the blower housing bolts.

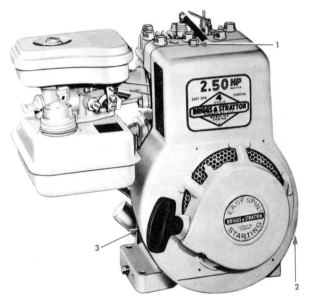

Figure 14-5. Most blower housings are held on with three bolts. (Briggs & Stratton Corp.)

Figure 14-7. Lift off the blower housing.

dirt like the ones in Figure 14-8? This will prevent air from passing through to take away the heat.

Use a scraper to clean the fins, as shown in Figure 14-9. Run the scraper between each of the

Figure 14-8. An engine with plugged cooling fins.

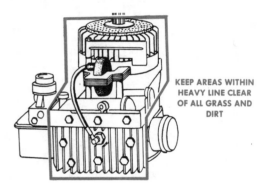

KEEP AREAS WITHIN
HEAVY LINE CLEAR
OF ALL GRASS AND
DIRT

Figure 14-10. Where cooling fins should be cleaned on a vertical engine. (Briggs & Stratton Corp.)

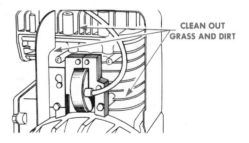

CLEAN OUT
GRASS AND DIRT

Figure 14-9. A scraper is used to clean the cooling fins on engine and flywheel.

Figure 14-11. Where cooling fins should be cleaned on a horizontal engine. (Briggs & Stratton Corp.)

cooling fins. Make sure the governor air vane can move. Move the vane back and forth with your hand. Blow away all the loosened dirt and grass. Figure 14-10 shows the area that should be cleaned on a vertical engine and, Figure 14-11, on a horizontal engine. When all the fins are cleaned, the blower housing may be put back on. Replace the three bolts or screws.

CLEANING THE AIR CLEANER

The air cleaner's job is to trap dirt before it can go into the engine. The longer the engine runs, or the dirtier the air, the more dirt is trapped in the air cleaner. The dirt can plug up the air cleaner. If air has a hard time getting through, the engine will be hard to start. It will not run well. The air cleaner must be cleaned at intervals recom-mended by the manufacturer so that it does not get plugged.

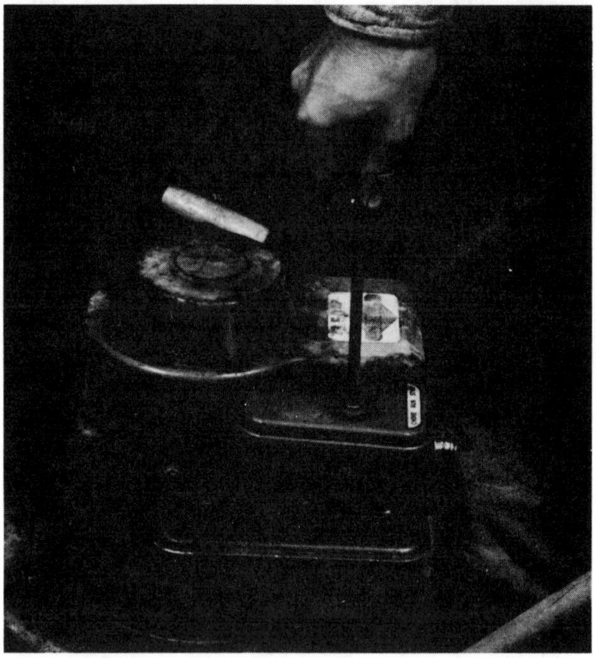

Air cleaners are mounted to the engine in one of two ways. The oil foam air cleaner is held on by a screw. The screw goes through the air cleaner and into the carburetor. Using a screwdriver, Figure 14-12, turn the screw to the left to remove it. Lift the air cleaner off the carburetor. Do it carefully so dirt does not fall into the carburetor.

Oil bath filters and many dry filters are held on by a nut. The nut fits on a shaft that sticks up from the carburetor. The nut is called a wing nut because it has small wings on it. Take off the wing nut. Use your hand to turn the wing nut to the left, Figure 14-13. Lift the air cleaner off the carbure-tor, Figure 14-14. Be careful that you do not let any dirt fall into the carburetor.

The oil foam air cleaner has three parts, as shown in Figure 14-15. Lift off the cover. The oil foam element may be taken out of the body. Wash the cover and body in a bucket of soapy water. Dry the parts with a rag.

Figure 14-12. Use a screwdriver to take out the air cleaner screw.

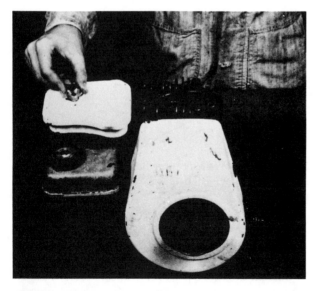

Figure 14-13. Remove the wing nut to remove the air cleaner.

Figure 14-14. Lift the air cleaner off carefully so dirt will not fall into the carburetor.

Figure 14-15. Remove the oil foam air cleaner by lifting off the cover. (Briggs & Stratton Corp.)

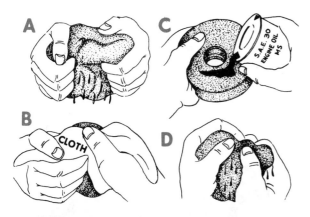

Figure 14-16. The steps used to clean an oil foam air cleaner element. (Briggs & Stratton Corp.)

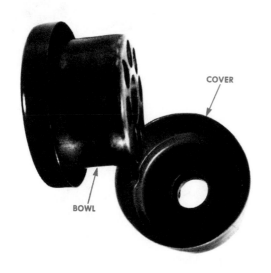

Figure 14-17. The oil is in the oil bath air cleaner bowl.

Follow the steps shown in Figure 14-16 to clean the oil foam element. Wash the foam in the soapy water. Wrap the foam in a dry rag and squeeze dry. Pour a small amount of engine oil on the foam. Squeeze the foam to get out most of the oil. Put the parts back together.

The oil bath air cleaner has two parts, a cover and a bowl, as shown in Figure 14-17. First lift the cover off as shown in Figure 14-18. The bowl is filled with dirty oil. Dump the oil out. Wipe the bowl clean with a rag. Place the bowl on a level bench. Pour clean oil back into the bowl. There is a line on the bowl marked "oil level." Fill the bowl to this level. Do not overfill or the oil will be sucked into the carburetor. Put the cover back on the bowl.

A dry paper air cleaner has two parts, a cover and a paper filter element. These parts are shown in Figure 14-19. Lift the cover off the filter element. Tap the element on the bench. This will make a lot of the dirt fall off. If the filter element is very dirty, it must be replaced with a new one.

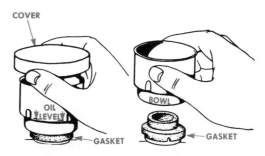

Figure 14-18. Lift the cover off the oil bath air cleaner bowl. (Briggs & Stratton Corp.)

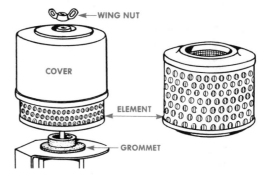

Figure 14-19. Remove the dry air cleaner element by lifting off the cover. (Briggs & Stratton Corp.)

Carefully place the air cleaner back on the carburetor. Tighten the screw or wing nut by turning to the right.

CLEANING AND INSPECTING CRANKCASE BREATHER

A crankcase breather is a valve, Figure 14-20, usually located on the side of the engine used to vent crankcase pressure. Compression pressure leakage around the piston rings enters the crankcase area. If this pressure is not vented, it could push oil out of the engine seals and gaskets. The breather consists of a small, round fiber disc. A small spring holds the disc closed. As crankcase pressure builds up, the disc is pushed off its seat and the pressure is vented. A tube often is connected from the breather assembly to the carburetor, Figure 14-21. This allows crankcase vapors to be burned in the engine.

After a period of time, the breather valve disc may get covered with sludge. The valve may stick open or closed. Periodically, the valve assembly must be removed from the engine and washed in solvent. The valve opening may be checked by inserting a wire feeler gage between the disc and

seat, as shown in Figure 14-22. A new gasket should be used when replacing the breather assembly.

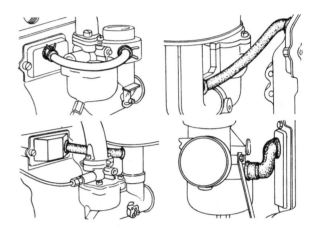

Figure 14-21. The breather is connected to the carburetor to burn the crankcase vapors. (Briggs & Stratton Corp.)

Figure 14-20. A crankcase breather vents crankcase pressures.

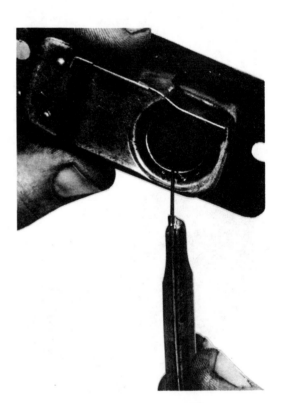

Figure 14-22. The breather valve opening is checked with a wire gage.

PREPARING AN ENGINE FOR STORAGE

Many times a small engine is used only during a certain season. A lawn mower may be used only during the spring and summer. A snowmobile may be used only during the winter months. When an engine is not going to be used for a period of months, it must be prepared for storage.

The first step in preparing an engine for storage is to remove all the fuel. Gasoline which is allowed to remain in a fuel system a long time turns to varnish. The varnish will plug carburetor passages and prevent proper carburetor operation. The fuel tank may be drained by removing the fuel line between the fuel tank and carburetor, Figure 14-23. Suction carburetor systems may be drained simply by turning the engine over with the fuel tank cap removed. The engine must be cold before draining the fuel to prevent a fire. The engine should then be started and run until the fuel in the carburetor is removed completely.

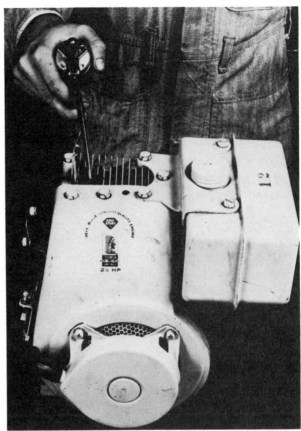

Figure 14-24. The cylinder is filled with oil to prevent the formation of rust.

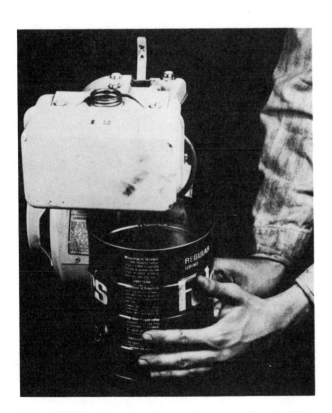

Figure 14-23. Draining fuel by disconnecting the fuel line to the carburetor.

During long periods of non-use, lubricating oil will run off the piston rings and cylinder walls and rust can form. The rust may cause the rings to stick to the cylinder wall and prevent the engine from turning over. The rust could cause the rings to score the cylinder wall when the engine is restarted.

To prevent rust, the cylinder should be filled with engine oil. Remove the spark plug and ground the spark plug wire. Turn the engine over until the piston is near the top of the compression stroke. Using a funnel, fill the combustion chamber area with oil, as shown in Figure 14-24. Replace the spark plug.

If the engine has a battery, it should be removed. The battery should be stored connected to a battery charger. Do not let the battery freeze. Store it in a warm, dry area on a board and not on concrete.

NEW TERMS

crankcase breather: A valve assembly used to vent crankcase pressure.

dry paper air cleaner: An air cleaner that uses a paper element to filter air.

oil bath air cleaner: An air cleaner that uses a container of oil to trap dirt going into the engine.

oil foam air cleaner: An air cleaner that uses a sponge-like foam element to trap dirt.

oil level: The amount of oil inside an engine.

SELF CHECK

1. What can happen if an engine does not have enough oil?
2. What can happen if an engine's oil gets too dirty?
3. How can you find out what kind of oil an engine uses?
4. Describe how to check oil level.
5. How do we drain old oil out of an engine?
6. What can happen to engine parts if cooling fins are plugged?
7. How do we clean an oil foam air cleaner?
8. How is oil used on an oil foam air cleaner?
9. How do we clean an oil bath air cleaner?
10. Describe how to prepare an engine for storage.

DISCUSSION TOPICS AND ACTIVITIES

1. Make a list of all the periodic maintenance jobs that should be done to a small engine you own.
2. Prepare a small engine for winter storage.

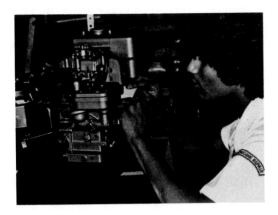

unit 15

tune-up: ignition system service

For best small-engine performance, the ignition and fuel systems must operate correctly. The engine requires a strong, clean, high-voltage spark at just the right time. The fuel system must provide exactly the right mixture of air and fuel for each of the engine's different operating ranges. Both the ignition and fuel systems require periodic adjustment and service in order to operate correctly. We call this series of service jobs a tuneup. In this unit we will see how to do the ignition service part of a tuneup on a small engine.

LET'S FIND OUT: When you finish reading and studying this unit, you should be able to:
1. Describe how to service spark plugs.
2. Explain how to remove, replace and adjust magneto breaker points.
3. Describe how to test the magneto coil and condenser.
4. Explain how to measure and adjust ignition timing.
5. Describe how to adjust the armature air gap.

SPARK PLUG SERVICE

The minute a spark plug is put in an engine, it begins to wear out. The center and ground electrodes are burned and wear away. Oil and gasoline build up on the electrodes. A dirty or worn-out spark plug may not allow the spark to jump across the electrodes. The engine may be hard to start, or may stop running. During a tuneup, the spark plug may be cleaned and regapped or it may be replaced with a new one.

Spark plugs are removed with a special deep socket wrench called a *spark plug socket wrench*, Figure 15-1. It is long enough to fit over the spark

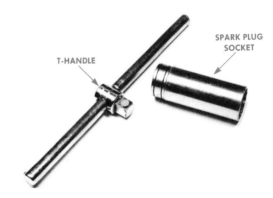

Figure 15-1. We use a spark plug socket and "T" handle to remove and replace spark plugs.

163

plug. The opening in the wrench, which is lined with a rubber insert, fits over the spark plug shell. The size is marked on the wrench. Most spark plugs use a 13/16" or 3/4" socket wrench. A handle called a "T" handle is used to turn the wrench.

First we must remove the spark plug from the engine. Remove the spark plug wire from the spark plug terminal and ground it for safety. Place the spark plug socket wrench, Figure 15-2, over the spark plug. Make sure it fits all the way over the spark plug. If the socket wrench is not down all the way, the spark plug could be broken. Remove the spark plug by turning it to the left.

Since the condition of each spark plug tells a story about the cylinder from which it was removed, it is a good idea to keep the plugs in order when working on a multiple-cylinder motorcycle or outboard. An easy method of doing this is to make a spark plug holder from a block of wood, Figure 15-3. Place the plugs in the holder with the terminal end down so they can be inspected without removing them from the holder.

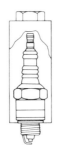

Figure 15-2. The wrench must fit all the way over the spark plug or the plug can be broken.

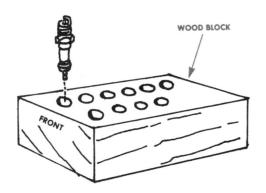

Figure 15-3. A spark plug holder can be made from a block of wood.

Inspect the appearance and condition of both ends of the insulator, the electrodes, gasket and shell of the plug. This will tell you how efficiently the engine has been operating and how suitable the spark plug is for the type of operation. Frequently the same type of spark plug used in two engines of the same make and model may show a big difference in appearance. The causes of these differences are the condition of the engine, its piston rings, carburetor setting, kind of fuel used and the conditions under which the engine is operated.

After removing the spark plug, examine the engine seat gasket of each plug. The gasket performs two important jobs. It conducts away much of the heat absorbed by the spark plug insulator tip from the burning fuel in the combustion chamber. Secondly, it maintains a gas-tight seal between the plug and its seat in the cylinder head. If a spark plug gasket is not tightly seated, leaking combustion gases will cause overheating of the plug.

If the gasket is flattened too much, the spark plug shell may be distorted or cracked and the plug gap may be changed by the excess torque. Figure 15-4 shows a spark plug gasket before installation, a gasket properly tightened, one not tightened enough and one tightened too much.

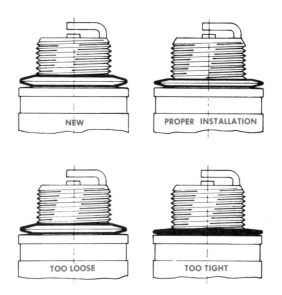

Figure 15-4. The spark plug gasket must be inspected. (Chevrolet Div.—General Motors Corp.)

A new spark plug gasket should be used every time a new or cleaned spark plug is installed for better performance and longer spark plug life.

Each spark plug should be compared to an analysis chart like that shown in Figure 15-5. Under normal operating conditions, spark plugs

NORMAL

Brown to grayish tan color and slight electrode wear. Correct heat range for engine and operating conditions.

RECOMMENDATION. Properly service and reinstall. Replace if over 10,000 miles of service.

SPLASHED DEPOSITS

Spotted deposits. Occurs shortly after long delayed tune-up. After a long period of misfiring, deposits may be loosened when normal combustion temperatures are restored by tune-up. During a high-speed run, these materials shed off the piston and head and are thrown against the hot insulator.

RECOMMENDATION. Clean and service the plugs properly and reinstall.

CARBON DEPOSITS

Dry soot.

RECOMMENDATION. Dry deposits indicate rich mixture or weak ignition. Check for clogged air cleaner, high float level, sticky choke or worn breaker contacts. Hotter plugs will temporarily provide additional fouling protection.

HIGH SPEED GLAZING

Insulator has yellowish, varnish-like color. Indicates combustion chamber temperatures have risen during hard, fast acceleration. Normal deposits do not get a chance to blow off, instead they melt to form a conductive coating.

RECOMMENDATION. If condition recurs, use plug type one step colder.

OIL DEPOSITS

Oily coating.

RECOMMENDATION. Caused by poor oil control. Oil is leaking past worn valve guides or piston rings into the combustion chamber. Hotter spark plug may temporarily relieve problem, but positive cure is to correct the condition with necessary repairs.

ASH DEPOSITS

Light brown to white colored deposits encrusted on the side or center electrodes or both. Derived from oil and/ or fuel additives. While non-conductive, excessive amounts may mask the spark, causing misfire.

RECOMMENDATION. If excessive deposits accumulate in short mileage, corrective measures may include installation of valve guide seals to prevent seepage of oil into combustion chamber.

TOO HOT

Blistered, white insulator, eroded electrodes and absence of deposits.

RECOMMENDATION. Check for correct plug heat range, overadvanced ignition timing, cooling system level and/or stoppages, lean fuel/air mixtures, leaking intake manifold, sticking valves, and if car is driven at high speeds most of the time.

MECHANICAL DAMAGE

Mechanical Damage to the plug's firing end is caused by some foreign object in the combustion chamber. It may also be due to the piston striking the firing tip of improper reach plugs. When working on an engine, be sure to keep the carburetor throat and any open plug holes covered. Consult the catalog for proper reach plugs.

Figure 15-5. Spark plug analysis chart. (Champion Spark Plug Co.)

wear out from intense heat, from the action of sulphur and lead compounds in the fuel, and from the erosion caused by the electric spark on the electrodes. Plugs that have been in operation for a long period of service should be replaced. Worn plugs cause loss of power, loss of speed, hard starting and sluggish performance.

Cleaning and Gapping Spark Plugs

Spark plugs coated with carbon or oxide deposits, Figure 15-6, should be cleaned in a sand or glass-bead blast cleaner. Follow the instructions of the cleaner manufacturer carefully. Plugs should be cleaned until the inside of the shell, the entire insulator and the electrodes are clean.

However, blast cleaning for too long will wear down the insulation and damage the plug. Plugs which cannot be completely cleaned with a reasonable amount of blasting should be replaced. Plugs with an oily or wet deposit should be cleaned in a nonflammable degreasing solution and thoroughly dried before blast cleaning to prevent the cleaning compound from packing into the shell. Spark plug service also includes cleaning the top insulator and terminal with a cleaning solvent to remove all oil and dirt.

After cleaning, inspect the plug carefully for cracks or other defects which may not have been

Figure 15-7. The electrode is filed straight with the spark plug file.

Figure 15-8. This is what a cleaned and filed spark plug looks like. (Champion Spark Plug Co.)

Figure 15-6. This is what a dirty worn spark plug looks like. (Champion Spark Plug Co.)

visible before. Look at the center electrode. The edge of the electrode is rounded. The spark does not like to jump off a rounded edge. We must use a spark plug file, Figure 15-7. Gently file the center electrode until it has a sharp edge like the one shown in Figure 15-8. Also, if blast cleaning has not removed all deposits from the electrodes, they should be cleaned with several strokes of fine abrasive paper or a file.

After cleaning the plug, look at the threads for carbon desposits. Clean them with a hand wire brush, taking care not to injure the electrodes or the insulator tip. Clean threads allow easier installation and proper seating of the plug when it is reinstalled in the engine. Plugs with badly nicked or damaged threads should be replaced.

Now we are ready to gap the spark plug. The gap or space between the two electrodes must be just right. You can look up what the gap should be in the repair manual for your engine. Many small engines use a gap of .030 (thirty thousandths of an inch).

Find the wire on the gapping tool, Figure 15-9, that is the same as the gap you want. Slide the wire into the space between the electrodes. It should just fit. If it will not go in, the gap is too small. If it goes in and is loose, the gap is too big.

The hook on the gapping tool is used to change the gap. The gap is changed by bending the ground electrode. To make the gap smaller, bend the ground electrode closer to the center electrode. To make it larger, bend the ground electrode away from the center electrode. After changing the gap, always recheck it. Never bend the center electrode; sidewise pressure on it may crack or break the insulator tip. Do not use gapping pliers, for they can damage the internal seals.

Now we are ready to put the spark plug back in

Figure 15-10. A torque wrench may be used to install a spark plug.

the engine. Screw the spark plug back into the cylinder head by hand. Turn it clockwise. Do not use the wrench until you cannot turn it any longer by hand. Then use the wrench to tighten the spark plug. Turn it about one-half turn. This should be enough. Do not overtighten it. If you do, you may damage the threads in the cylinder head. Some engines have torque specifications for the spark plug. A torque wrench, Figure 15-10, can be used to install the spark plug to the correct torque setting.

MAGNETO SERVICE

The main parts of the magneto ignition usually are located under the engine's flywheel. The first step in magneto service is to remove the flywheel. On a moped or motorcycle engine, a side cover must be removed to get at the armature or flywheel assembly. The cover on an outboard engine power head is removed to get at the flywheel. Most small engines require that the fan housing be removed to work on the flywheel.

The flywheel is driven by the crankshaft. The end of the crankshaft fits into a tapered hole in the middle of the flywheel. A key is positioned in a keyway and keyseat between the flywheel and crankshaft to provide a positive lock. The end of the crankshaft is threaded. A large nut or part of a rewind starter is threaded onto the crankshaft to hold the flywheel in position.

Figure 15-9. The gap is measured and adjusted with a gapping tool.

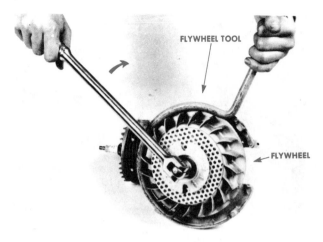

Figure 15-11. A flywheel tool is used to hold the flywheel while removing the nut.

Figure 15-12. A rewind starter clutch is removed with a flywheel holder and starter wrench.

Figure 15-13. Removing a flywheel by striking an impact nut.

Flywheels may be removed in several ways. First, the nut or rewind clutch must be removed. In order to remove the nut, you must keep the flywheel from rotating. A flywheel holder, Figure 15-11, prevents the flywheel from turning. Many engines have a rope starter that rewinds itself after pulling. A starter clutch often is used with this arrangement. The clutch threads onto the crankshaft. The clutch must be removed to remove the flywheel. A starter clutch wrench must be used to remove the starter, as shown in Figure 15-12.

Most rewind starters have right-hand threads. Many flywheel nuts, however, have left-hand threads. Make sure the nut is loosened by turning in the correct direction. Closely inspect the visible threads or check a service manual.

With the nut removed, the flywheel may be removed from the crankshaft. The flywheel is locked to the crankshaft by the tapered fit between the two. If the tapered fit is exceptionally tight, or if there is a corrosion buildup between the flywheel and crankshaft, the flywheel may be difficult to remove. A large nut called an impact nut can be threaded onto the crankshaft. This will protect the threads on the shaft. Insert a large pry bar under the flywheel as shown in Figure 15-13, and, while prying upward with the pry bar, tap squarely on the impact nut to break the flywheel loose. If necessary, rotate the flywheel a half turn and repeat the process until the flywheel loosens from the crankshaft. Remove the nut and flywheel. Never hammer directly on the crankshaft or crankshaft nut because this would damage the crankshaft threads.

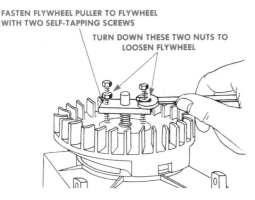

FASTEN FLYWHEEL PULLER TO FLYWHEEL
WITH TWO SELF-TAPPING SCREWS

TURN DOWN THESE TWO NUTS TO
LOOSEN FLYWHEEL

Figure 15-14. Removing a flywheel with a puller.

Figure 15-15. An automotive puller like this can break a flywheel.

If this procedure fails to loosen the flywheel, the next step is to use a puller. Many flywheels have holes near the hub. These are used with a puller. Bolts from the puller are threaded into the flywheel holes. The center part of the puller is then threaded down to push against the crankshaft, Figure 15-14. This forces the flywheel off the crankshaft.

Do not use an automotive style puller, Figure 15-15. This kind of puller grips the outside rim of the flywheel. The flywheel is very thin in the center. Very little pressure from this kind of puller will bend or even break the flywheel.

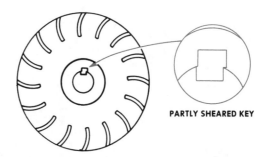

PARTLY SHEARED KEY

Figure 15-16. The flywheel is inspected for a loose or partly sheared key. (Briggs & Stratton Corp.)

Flywheel Inspection

The flywheel is an important component of the magneto ignition system. It should be inspected carefully. It is possible that the magnets in the flywheel may lose their magnetism. Place an unmagnetized screwdriver near the magnet. It should be strongly attracted to each magnet in the flywheel. Some older flywheels have replaceable magnets but most are cast into the flywheel. A magnet that has lost magnetism must be replaced either by replacing the individual magnet or by replacing the flywheel.

Another critical area for inspection is the key and keyseat area. The key positions the flywheel in relation to the crankshaft. This in turn determines when ignition will occur. A loose-fitting key can cause the relationship between the crankshaft and flywheel to change and, in turn, change ignition timing. Many lawn mower engines use a key made from a soft material. If the lawn mower blades strike an object, the key will shear. This prevents the flywheel momentum from breaking the crankshaft. Once the key is partially sheared, the timing will be affected and the engine probably will not run. Never replace these special soft keys with a hardened steel key. A loose key and a partially sheared key are shown in Figure 15-16. Also check the keyseat in an aluminum flywheel for cracks and replace the flywheel if cracks are found.

Replacing Breaker Points and Condenser

The breaker points act as the switch that triggers the high-voltage buildup in the magneto

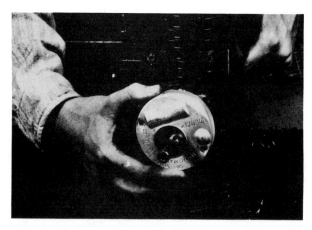

Figure 15-17. The breaker point cover must be removed to service the points.

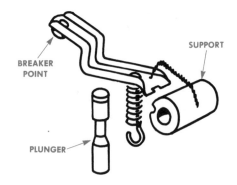

Figure 15-18. Contact points operated by a plunger. (Briggs & Stratton Corp.)

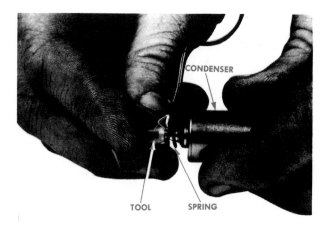

Figure 15-19. Removing primary wire from the condenser.

coil. After a period of time, current flow across the breaker points breaks down the contact surfaces. The destruction of the contact surfaces can become severe enough to affect current buildup in the coil. The engine begins to lose power and waste gasoline. If the contact points are not serviced, they may get bad enough so the engine stops running altogether. For these reasons, the breaker points are changed periodically.

The condenser's job is to protect the breaker points and improve magneto coil operation, by soaking up excess current that tries to jump across open breaker points. If the condenser fails, the engine could fail to run. To prevent condenser failure, a new condenser is installed each time the breaker points are replaced.

On most small engines, the breaker points are located behind a cover, Figure 15-17. The breaker point cover protects the points from dirt and water. The primary wires go through the cover. A sealer is used where the wires go through the cover to prevent moisture from getting on the points. The breaker point cover must be removed to service the points. The cover is removed by removing two hold-down screws or a hold-down clip.

Two general styles of breaker points are used on small engines. One type uses a plunger operated by a flat spot on the crankshaft to operate the movable breaker point, Figure 15-18. The condenser and stationary breaker point are combined into one unit.

Replace this type of breaker point set by removing the hold-down screw through the movable breaker point. First unhook the spring from the movable breaker arm. Then remove the stationary contact and condenser by removing the screw holding the condenser clamp to the engine. The primary wire is attached to the condenser. The wire is held in place by a small spring. Use the small depresser tool supplied with the new condenser to push down on the spring and remove the wire, as shown in Figure 15-19.

Install the new breaker point set attaching the primary wire to the new condenser, using the depressing tool. The new condenser is then mounted to the engine. Position the new movable breaker point and attach the spring.

The gap between the points must be set accurately since it determines coil buildup time and sets engine timing. Turn the crankshaft until the contact points are open to their widest gap. Use a flat feeler gage to measure the gap. Select the feeler gage of the recommended size and position it between the breaker points, as shown in Figure 15-20. The gap is adjusted by loosening the lock screw holding down the condenser and moving the condenser up and down. The gap is adjusted properly when the feeler gage will slide in and out of the gap with a light drag.

Another common style of breaker points is shown in Figure 15-21. These points are operated by a removable cam attached to the crankshaft. The condenser is separate from the breaker

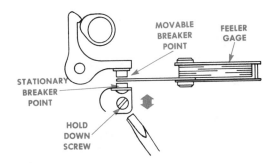

Figure 15-22. The breaker points are adjusted by moving the stationary point.

points. A dust cover held by a spring cup covers the breaker point assembly. To remove the points first remove the spring clip and dust cover. Disconnect the primary wire and condenser wire. The movable point is removed by taking off the lock ring on the mounting stud. The point assembly may be lifted off the stud. Remove the screw that holds the stationary point and lift it off the engine. Remove the condenser by removing the hold-down screw through the condenser clamp.

Install the new breaker points by placing the new movable point on the mounting stud and installing the lock ring. Position the stationary point and replace the hold-down screw. Install the new condenser and reconnect the primary and condenser wire.

The engine is rotated until the high part of the cam contacts the movable breaker point arm, as shown in Figure 15-22. The correct size flat feeler gage is placed between the breaker points. The gap is adjusted by loosening the stationary point hold-down screw and sliding the stationary point back and forth. The gap is correct when the feeler gage fits through the gap with a light drag.

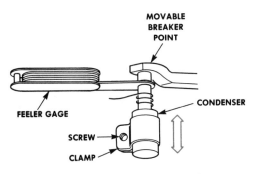

Figure 15-20. Adjusting the breaker points by moving the condenser.

Testing the Magneto Coil

Inspect coil assembly for damage that may affect its operation. Look especially for cracks or gouges in insulation, evidence of overheating or other damage. Make sure electrical leads are intact, especially where they enter the coil.

With coil assembly installed on core assembly, check the operation of the coil on an approved

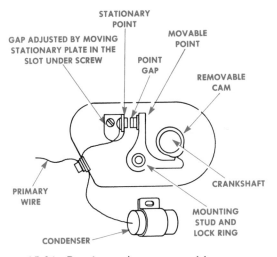

Figure 15-21. Breaker points operated by a cam.

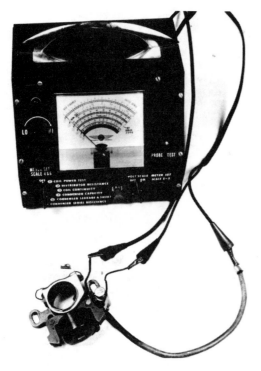

Figure 15-23. Testing the magneto coil for power.

Figure 15-24. A probe is used to test the magneto coil insulation.

coil tester, similar to that shown in Figure 15-23. When the coil is in good condition, it will produce a spark across the gap at the recommended dial setting. Refer to the specification chart supplied by the test equipment manufacturer and use the appropriate control setting for the coil assembly being tested. If the coil fails to produce a continuous spark at this setting, the coil should be replaced.

Check outer insulation of the coil for high-tension leaks. Use the probe provided with the test equipment, Figure 15-24. Move the probe slowly around the coil, holding the probe 1/8 inch from the insulation. If a weak spot is present in the insulation, a spark will jump from coil to probe. Replace a coil that has weak insulation.

If a coil tester is not available, substitute a new coil for the old and try out the engine. If it operates properly, the coil was at fault.

Testing the Condenser

Both new and old condensers should be tested when the ignition system is serviced. Sometimes a new condenser will not test as well as the old one. In this case the old condenser should be reused.

Figure 15-25. Testing the condenser.

Inspect the condenser for visible damage. Look especially for damaged terminal lead, dents or gouges in can or broken mounting clip.

Check the condenser on a high-quality tester, Figure 15-25. Follow the test equipment manufacturer's instructions to check for capacitance and series resistance. Capacitance is measured in microfarads, abbreviated mfd. A condenser in good condition should have a capacitance of around .17 to .23 mfd. If the capacitance is too high or too low when compared to manufacturer's specifications, the breaker points will be worn away rapidly.

SETTING IGNITION TIMING

Many small engines and all engines used on mopeds, motorcycles and outboards have an ignition timing adjustment. For the best engine performance, the spark must be introduced in the engine's combustion chamber at exactly the correct time in relation to piston position.

The time that the spark is introduced into the cylinder is measured by the position of the crankshaft. When the piston is at top dead center, the crankshaft and connecting rod are lined up vertically. The crankshaft is at an angle to this vertical line when the piston is at a point before or after top dead center, Figure 15-26. This angle, measured in degrees, is used to describe when ignition occurs. The letters BTDC commonly are used to describe angles *before top dead center* and ATDC to specify angles *after top dead center*.

The time at which ignition occurs in relation to crankshaft movement is measured by measuring the position of the piston in relation to top dead center, Figure 15-26. The manufacturer provides specifications for this measurement. The measurement is easiest to make with the cylinder head removed. Turn the engine over by hand until the piston is up at the top on the engine's compression stroke. Then turn the engine backwards enough to lower the piston in the cylinder. A scale or dial indicator may be used to move the piston into the recommended position.

With this distance set, the magneto must be adjusted so that it is ready to provide ignition. Magneto timing is adjusted by moving the mag-

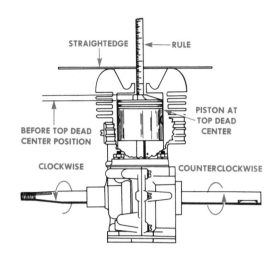

Figure 15-26. The ignition timing is measured with the piston before top dead center. (Tecumseh Products)

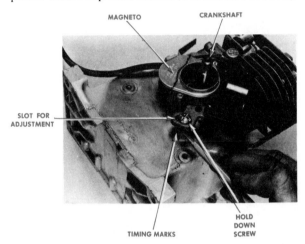

Figure 15-27. The magneto is timed by moving it in relation to the crankshaft.

neto assembly in relation to the crankshaft, Figure 15-27. The magneto is mounted to the engine with several hold-down bolts. The holes in the magneto are slotted to allow the magneto to be moved back and forth. Moving the magneto moves the breaker points in relation to the cam on the crankshaft. Moving the points in a direction opposite to cam rotation will cause the timing to occur earlier. Moving the points in the same direction as the cam turns will cause ignition to occur later.

When the piston is in the desired position, rotate the magneto to a position where the contact points just start to open. When the points open, the spark occurs.

If a magneto is removed from an engine, reference marks should be scribed on the magneto and engine. This will prevent having to retime the engine. Anytime the breaker point spacing is changed or adjusted, the timing also must be adjusted.

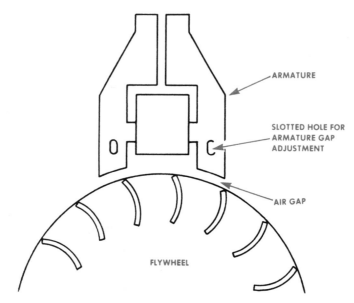

Figure 15-28. The armature gap is adjusted with slotted holes.

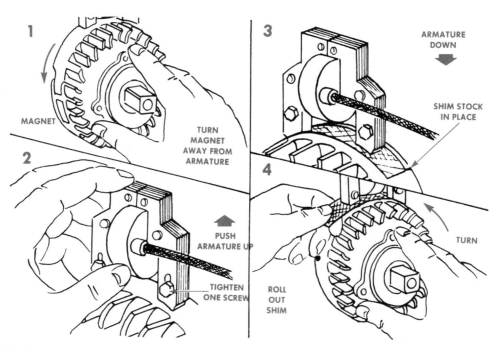

Figure 15-29. Setting the armature gap. (Briggs & Stratton Corp.)

INSTALLING THE FLYWHEEL

When the breaker points have been serviced and the timing adjusted, the flywheel may be installed. Always use the correct key between the crankshaft and flywheel. Install the key and push the flywheel on the crankshaft. Use a crankshaft holder to prevent the flywheel from turning. Install the flywheel nut or starter rewind clutch. Always use a torque wrench to tighten the nut.

SETTING THE ARMATURE GAP

The armature gap or "E" gap is the space between the magneto coil armature and the flywheel magnets. On many magnetos this space is adjustable. The gap must be correct. If the space is too small, the magnets and armature may contact each other and wear rapidly. If the space is too large, the magnetic field will not be strong enough to build primary current.

The armature is attached to the engine with several hold-down screws. The holes in the armature are slotted so the armature can be moved up and down, as shown in Figure 15-28. The gap measurement is specified by the manufacturer. Rotate the flywheel, Figure 15-29, until the magnets are under the armature. Place a nonmetallic piece of paper or cardboard shim the thickness of a business card between the armature and flywheel. Loosen the armature and push it down to touch the paper. Then tighten the armature and remove the paper. Always rotate the flywheel several times to be sure there are no high spots that could contact the armature.

NEW TERMS

armature gap: The space between the armature and flywheel magnets.

breaker point gap: The space between the movable and stationary points when they are open.

spark plug analysis chart: A chart showing common problems of spark plugs.

spark plug gap: The space between the center and ground electrode of a spark plug.

spark plug socket: A special socket made to remove spark plugs without damaging them.

tune-up: The replacing of worn ignition parts and adjustment of other systems to get the best performance from an engine.

SELF CHECK

1. What will a dirty or worn-out spark plug do to an engine?
2. For what is a spark plug analysis chart used?
3. Explain how to clean a spark plug.
4. Why is the center electrode filed?
5. Explain how to measure a spark plug gap.
6. How do you make the spark plug gap smaller? Larger?
7. Describe how to change a set of breaker points.
8. Describe how to adjust contact points.
9. Explain how to set ignition timing.
10. How is armature air gap adjusted?

DISCUSSION TOPICS AND ACTIVITIES

1. Remove a spark plug from an engine and compare it to an analysis chart. What problems can you identify?
2. Remove and replace a set of contact points from a shop engine.

unit 16

tune up: fuel system service

Servicing the fuel system is an important part of every tune-up. The mechanic normally services the ignition system first, then the fuel system. The mechanic will need to determine how much fuel-system service is necessary. Often just adjusting the carburetor is all that is necessary. If the fuel-system components have not been serviced in a long time or if they have been used with dirty fuel, major fuel-system service will be necessary. In this unit we will study how to service the complete fuel system.

LET'S FIND OUT: **When you finish reading and studying this unit, you should be able to:**
1. **Describe how to clean fuel tanks and fuel filters.**
2. **Explain how to service a suction carburetor.**
3. **Describe how to service a float carburetor.**
4. **Explain how to service a diaphragm carburetor.**
5. **List the steps to follow in adjusting a carburetor.**

FUEL TANK SERVICE

The small engine fuel tank usually causes very little trouble. If the fuel is allowed to stay in the tank for a long time it can become stale. Old fuel should be removed from the tank by opening the drain at the bottom. If there is no drain and the engine cannot be turned over, you will have to remove the tank.

After draining, the tank should be flushed with fresh cleaning solvent to wash out any stale fuel. Check the inside of the tank for signs of rust. A tank on an engine left outdoors for a long time may have rust buildup. Unless the rust is com-

BALL BEARINGS

Figure 16-1. Filling the fuel tank with ball bearings and shaking the tank will remove rust.

pletely removed, it will flake off and cause problems with the carburetor.

One way to remove rust in a tank is to fill the tank with small ball bearings, Figure 16-1. As the tank is shaken up and down, the ball bearings will knock off the rust. The tank is then flushed with clean solvent.

The filler cap, Figure 16-2, has a tiny hole in the center that acts as a vent. Atmospheric pressure must be able to get into the tank or fuel will not move into the carburetor. The vent often becomes clogged. Compressed air may be blown through the vent to clear it.

Figure 16-2. The fuel tank lid has a vent hole that must be kept open.

FUEL FILTER SERVICE

Many small engines have a fuel filter, Figure 16-3, between the fuel tank and carburetor. This filter may become clogged as it filters dirt from the gasoline. The filter may be a fine mesh screen or a bronze element. Often it is attached to the

Figure 16-3. A clogged fuel filter must be cleaned or replaced.

Figure 16-4. A sediment bowl is serviced by removing and cleaning it.

bottom of the fuel tank where the fuel line is connected. Check the filter by removing it from the tank. If it is damaged or clogged, it should be cleaned or replaced. Filters may be washed in clean solvent and blown dry with compressed air. Always check a filter by blowing on the fuel inlet end. Air should pass freely through the filter element.

Some engines use a glass bowl called a *sediment bowl* attached to the fuel tank, Figure 16-4. Fuel passes through the bowl on the way to the carburetor. Dirt in the fuel drops to the bottom of the sediment bowl. The bowl is made of glass so the level of sediment is visible. When the fuel system is serviced, the bowl should be cleaned. A retaining nut on the bottom of the bowl is loosened, allowing the bowl to be removed. A gasket and filter screen located above the bowl should be removed also. When you replace the cleaned bowl, also install a new gasket.

CARBURETOR REMOVAL AND CLEANING

When troubleshooting indicates a problem with the carburetor, it must be removed from the engine and serviced. Major carburetor service must be done with the carburetor off the engine. Remove the air cleaner and gasket. Disconnect the fuel line from the carburetor. Disconnect the governor accelerator linkage. Remove nuts and/or bolts, gasket and carburetor.

Locate and follow the right shop manual when disassembling a carburetor. The carburetor is disassembled according to a set procedure. When disassembled, clean the major components in a cold tank filled with carburetor cleaner. Wear a face shield and rubber gloves when using the cold tank. Rubber parts, plastic parts and diaphragms should not be put in carburetor cleaner. Blow out all passages in castings with compressed air. Do not pass drills through jets or passages. In the following sections we will study the service procedures used on specific types of carburetors.

Suction Carburetor Service

Remove the carburetor and fuel tank as one unit, being careful not to bend the governor linkage. On models equipped with a stop switch, remove the ground wire. After removal of the carburetor from the fuel tank, inspect the tank for deposits of dirt and/or stale fuel. The tank should be cleaned in solvent.

Cast throttles, Figure 16-5, are removed by backing off the idle-speed adjusting screw until the throttle clears the retaining lug on the carburetor body. Stamped throttles are removed by using a phillips screwdriver to remove the throttle valve screw. After removal of the valve, the throttle may be lifted out with a pencil. To replace the throttle after cleaning, just reverse the disassembly procedure.

Figure 16-6. A nylon fuel pipe is removed with a socket wrench.

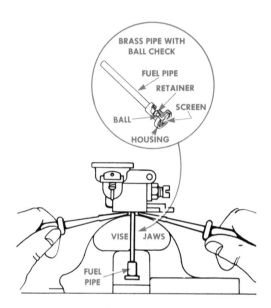

Figure 16-7. A brass fuel pipe is removed by clamping it in a vise and prying it out (Briggs & Stratton Corp.)

The fuel pipe contains a check ball and a fine mesh screen. To function properly, the screen must be clean and the check ball free. Replace the pipe if the screen and ball cannot be cleaned satisfactorily in carburetor cleaner. **DO NOT LEAVE CARBURETOR IN CLEANER MORE THAN ½ HOUR WITHOUT REMOVING NYLON PARTS.** Remove nylon fuel pipes and replace them using a socket wrench, as shown in Figure 16-6. Remove brass fuel pipes by clamping the pipe in a vise and prying it out, as shown in Figure 16-7.

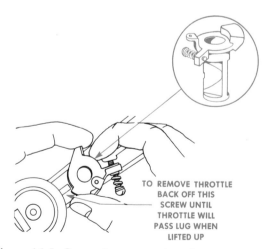

TO REMOVE THROTTLE BACK OFF THIS SCREW UNTIL THROTTLE WILL PASS LUG WHEN LIFTED UP

Figure 16-5. Removing a cast throttle.

To install brass fuel pipes, remove the throttle, if necessary, and place the carburetor and pipe in a vise. Press the pipe into the carburetor until it projects from the carburetor face as shown in Figure 16-8.

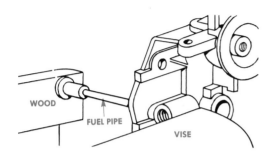

Figure 16-8. The brass fuel pipe is pressed back in with a vise. (Briggs & Stratton Corp.)

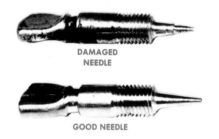

Figure 16-9. Comparison of a good and a damaged needle.

Remove the needle valve and spring by unscrewing them. Carefully inspect the tapered end. If there are any nicks, scores or ridge lines on the end of the needle, Figure 16-9, you should replace it. A damaged needle will not allow the mixture to be controlled properly. The needle seat is removed from the side of the carburetor with a wrench. Clean the seat with carburetor cleaner. If the needle requires replacement, change the seat at the same time. Replace the seat with a new gasket. The needle is replaced in the seat and adjusted as explained in another section. After you clean and install the new parts, mount the carburetor to the fuel tank. The fuel tank and carburetor assembly then are mounted to the engine.

Float Carburetor Service

The float carburetor is removed by removing the two cap screws which attach it to the engine. Disconnect the fuel line, throttle and governor linkage. Remove idle mixture adjusting needle valve by unscrewing it. Then loosen the high-speed needle valve packing nut. Remove the packing nut and high-speed needle valve together. To remove the nozzle, use a narrow, blunt screwdriver so as not to damage threads in the lower carburetor body. The nozzle projects diagonally into a recess in the upper body and must be removed before the upper body is separated from the lower body, or it may be damaged. (See Figure 16-10.) Remove the screws holding the upper

Figure 16-10. Parts of a float carburetor.

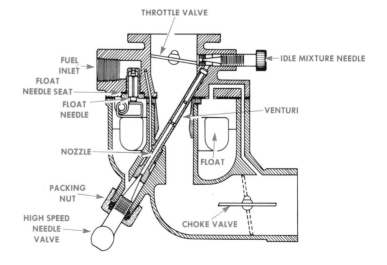

Figure 16-11. Checking for a parallel float.

Figure 16-12. Adjusting the float by bending the tang.

old gaskets harden and may leak. Tighten the inlet seat with the gasket securely in place, if used. Some float valves have a spring clip to connect the float valve to the float tang. Other valves are of nylon with a clip which fits over the float.

With the body gasket in place on the upper body and the float valve and float installed, the float should be parallel to the body mounting surface, Figure 16-11. If it is not, bend the tang on the float, Figure 16-12, until it is parallel to the surface. DO NOT PRESS ON FLOAT TO ADJUST. Some manufacturers have a specific float adjustment measurement. The distance from the body to the float is measured with a scale or drill, as shown in Figure 16-13.

Assemble venturi and venturi gasket to lower body. Be sure holes in the venturi and venturi gasket are aligned. Some models do not have a removable venturi. Install choke parts and welch plug if previously removed. Use a sealer around the welch plug to prevent entry of dirt.

Fasten upper and lower bodies together with the mounting screws. Screw in the nozzle with a narrow, blunt screwdriver, being careful that the nozzle tip enters the recess in the upper body. Tighten nozzle securely. Screw in the needle valve and idle valve until they just seat. Back off the needle valve 1½ turns. Do not overtighten the packing nut. Back off the idle valve ¾ turn. These settings are approximately correct. Final adjustment will be made when the engine is running as explained in a later section.

and lower bodies. A pin holds the float in place. Remove the pin to take out the float and float valve needle. Check the float for leakage. If it contains gasoline or is crushed, it must be replaced. Use a wide, properly fitting screwdriver to remove the float inlet seat, if used. Lift the venturi (if removable) out of the lower body of a small engine carburetor. Some carburetors have a welch plug. This should be removed only if necessary to remove the choke plate. Some carburetors have a nylon choke shaft.

Use new parts where necessary. Carburetor repair kits are available. Always use new gaskets;

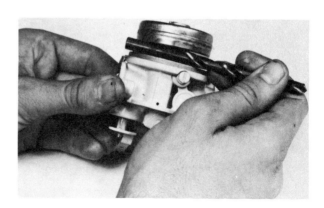

Figure 16-13. Measuring the float setting.

Diaphragm Carburetor Service

Before you remove the carburetor, the outside must be cleaned carefully to prevent dirt from entering the crankcase. Remove the fuel line, throttle control and hold-down bolts. Remove the carburetor. An exploded view, Figure 16-14, is helpful in servicing the carburetor.

Remove the pump diaphragm cover screws and cover. Then remove pump gasket and pump diaphragm. Dislodge the filtering screen. Remove main diaphragm cover screws and cover, main diaphragm and gasket. Take off the fulcrum pin screw, pin, control lever and spring. Finally remove the inlet needle.

Cold tank cleaner can be used on all parts except diaphragms and gaskets. Before reassembling, rinse all parts in clean solvent and blow with compressed air. Do not use cloth, because tiny particles of lint adhering to carburetor parts

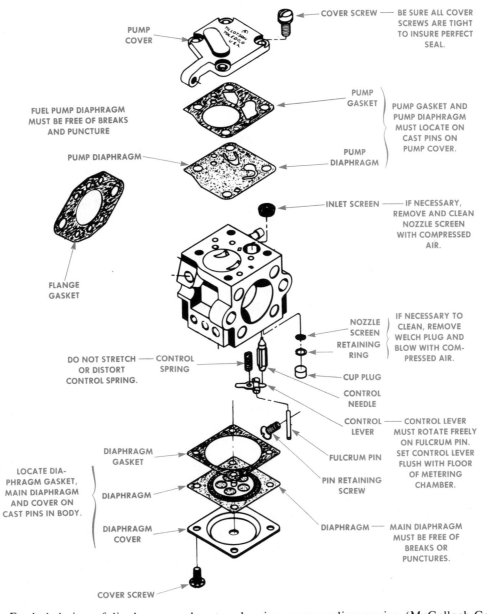

Figure 16-14. Exploded view of diaphragm carburetor showing areas needing service. (McCulloch Corp.)

LOW

Figure 16-15. Adjusting a low inlet control lever. (McCulloch Corp.)

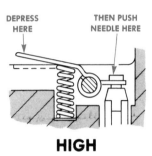

HIGH

Figure 16-16. Adjusting a high inlet control lever. (McCulloch Corp.)

will cause a problem. Channels in the metering body can be cleaned by blowing through idle- and high-speed mixture screw openings. Do not use wires or drills to clean orifices.

Be certain the diaphragm or diaphragms are installed correctly. On the metering body install the pump diaphragm, then the gasket, the main diaphragm gasket and the main diaphragm. Tighten all screws evenly to insure a complete seal.

When reassembling the inlet control lever and spring, take care to see that the spring rests in the well of the metering body and locates on the dimple of the inlet control lever. Do not stretch the spring. The inlet control lever is set properly when it is flush with the floor of the diaphragm chamber. If the diaphragm end of the control lever is low, pry it up as shown in Figure 16-15. If the lever is high, depress the diaphragm end and push on the needle for proper adjustment, as shown in Figure 16-16. On some models, the inlet control lever is hooked to the inlet needle at one end. Special care is required when reassembling these parts to insure proper operation.

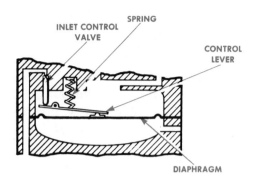

Figure 16-17. Inlet control lever and control valve assembly. (McCulloch Corp.)

The carburetor inlet control valve, Figure 16-17, regulates the amount of fuel which can be pumped into the carburetor by the fuel pump. The valve is operated by the diaphragm, diaphragm spring and control lever so that each of these can affect the air-fuel ratio.

Dirt in the fuel is the principal cause of trouble for the inlet control valve. Dirt caught between the inlet valve and its seat can prevent the valve from closing completely. This results in a continuous flow of fuel from the fuel pump through the fuel chamber and into the carburetor air passage and the engine is flooded.

On the other hand, if the diaphragm spring is compressed, the inlet valve may close all the way, allowing the fuel pump to force fuel through the carburetor into the passage. This extra fuel will flood the engine. If the spring is stretched, it will prevent proper adjustment of the carburetor and the mixture will tend to be too lean.

Install the reassembled carburetor on the engine. Tighten the low- and high-speed adjustment needles finger-tight, then back them out one to two turns. This provides a preliminary start-up adjustment that must be final-adjusted later when the engine is running.

CARBURETOR ADJUSTMENT

An engine needs just the right amount of air-fuel mixture. Too much or too little will cause the engine to run roughly. After you service the carburetor, you must adjust the air-fuel mixture.

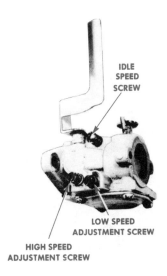

Figure 16-18. Mixture-adjusting screws on a vacuum carburetor. (Briggs & Stratton Corp.)

Figure 16-20. Mixture-adjusting screws on a diaphragm carburetor.

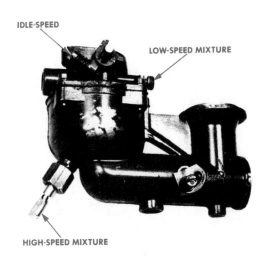

Figure 16-19. Adjusting screws on a float carburetor.

The air-fuel mixture is adjusted by mixture adjustment screws. Many carburetors used on small engines have a high-speed mixture screw. This screw is used to adjust the carburetor's mixture at high speed. Many small engines also have a low-speed mixture adjusting screw. This screw is used to adjust the carburetor's mixture at low

speed. Most carburetors have an idle-speed screw as well. This screw is used to set the speed of the engine when the throttle valve is closed.

The suction carburetor, shown in Figure 16-18, has two adjusting screws. The high-speed, fuel-mixture screw controls the fuel coming up the fuel pipe. An idle-speed screw controls how far the throttle valve can close. The idle speed screw is used to set the engine's idle speed.

The float carburetor, shown in Figure 16-19, has three adjusting screws. It has a high-speed and a low-speed fuel-mixture screw. It also has an idle-speed adjusting screw. A diaphragm carburetor is shown in Figure 16-20. It has the same three adjusting screws as the float carburetor. All types of carburetors are adjusted in the same way. Next we will see how to make these adjustments.

Adjusting the High-Speed Mixture

Start up the engine. Allow it to run for several minutes to warm up. Find the carburetor high-speed mixture screw, Figure 16-21. Some screws have a slot for a screwdriver. Others have a handle; these can be adjusted by hand.

Open the engine's throttle so that it runs at the speed specified by the manufacturer. Turn the

Figure 16-21. Adjusting an engine's high-speed mixture.

screw in or clockwise. This closes off the mixture to the venturi. Turn the screw until the engine runs roughly. Hold the throttle steady. Turn the screw out slowly. The engine will begin to speed up. If you turn it too far, the engine will begin to slow down. Stop turning when the engine runs as fast as it will run.

Adjusting the Low-Speed Mixture

Close the throttle. Allow the engine to idle. Find the idle mixture-adjustment screw, Figure 16-22.

Put the screwdriver into the idle mixture-adjusting screw. Turn the screw in or clockwise. This shuts off the low-speed mixture. Turn the screw slowly. Stop turning when the engine begins to run roughly. Turn the screw out or counterclockwise. Turn it slowly. This allows more mixture into the engine. The engine will begin to speed up. If you go too far, the engine will begin to

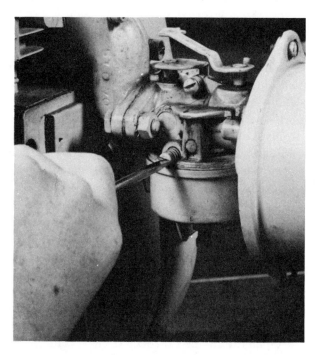

Figure 16-22. Adjusting an engine's low-speed mixture.

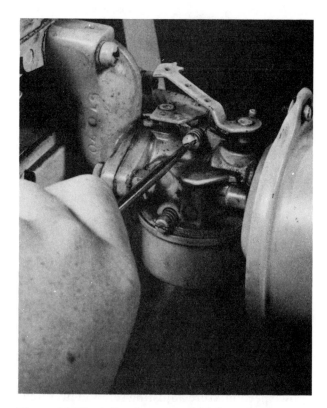

Figure 16-23. Adjusting an engine's idle speed.

slow down. Stop turning the screw when the engine runs as fast as it will run.

Adjusting Idle Speed

After the mixture adjustments are complete, the idle speed may be adjusted. The engine manufacturer specifies a certain idle speed in RPM. Find the idle speed specifications. Idle speed is measured with an electronic instrument called a *tachometer*. The tachometer is connected to the magneto ignition and counts the ignition impulses. The scale on the tachometer is divided into revolutions per minute (RPM). Connect the tachometer to the engine.

Start up the engine. Allow it to idle for several minutes and warm up. Find the idle-speed screw. Put the screwdriver into the idle-speed screw, Figure 16-23. Turn the screw to the right. This opens the throttle valve. The engine will begin to run faster. The engine should run just fast enough so that it will not stop. If it is running too fast, turn the screw to the left. This closes the throttle valve. The engine will slow down. Set the idle to specifications.

NEW TERMS

carburetor service: The removal, disassembly, cleaning and reassembly of a carburetor.

float adjustment: Setting the float level in the carburetor by bending the float tang.

high-speed adjustment: The setting of a carburetor's air-fuel mixture at high speed.

idle-speed adjustment: The setting of a carburetor's idle speed or RPM when the machine is not in motion.

low-speed adjustment: The setting of a carburetor's air-fuel mixture at low speed.

tachometer: An electronic instrument used to measure engine speed or RPM.

tang: A small tap on a float that contacts the needle valve.

SELF CHECK

1. Describe how to clean out a rusty fuel tank.
2. Why should fuel filters be cleaned?
3. Describe the procedure used to service a suction carburetor.
4. Explain how to service a float carburetor.
5. How is the float adjusted?
6. Explain how to service a diaphragm carburetor.
7. What adjustment screws does a suction carburetor have?
8. What adjustment screws does a float carburetor have?
9. What adjustment screws does a diaphragm carburetor have?
10. Describe how to adjust a carburetor for high speed, low speed and idle.

DISCUSSION TOPICS AND ACTIVITIES

1. Use a shop carburetor to practice disassembly procedures.
2. Practice adjusting a carburetor on an engine that runs.

Experimental and home-built
aircraft use small engines.

186

part 6
engine
service

When a small engine has a major problem inside or when it is worn out, it must be overhauled. Overhauling involves several steps. First, the engine is disassembled and cleaned. Each of the parts is inspected and measured for wear. The individual components are then reconditioned. The engine may then be reassembled again. We will study all the techniques used in engine reconditioning.

unit 17

engine disassembly and cleaning

When troubleshooting indicates a major problem, the engine must be disassembled for service. After the engine is disassembled, the next step is to clean each of the components thoroughly. Parts must be cleaned so that the mechanic can make a diagnosis of problems and measure for wear. Dirt and other foreign material is the number one enemy of the engine. Engine parts must be cleaned and kept clean throughout the rebuilding operation. In this unit we will see how the engine is disassembled and cleaned.

LET'S FIND OUT: When you finish reading and studying this unit, you should be able to:
1. Describe the procedure for removing valves from an engine.
2. Explain the procedure for removing a cylinder ridge.
3. List the steps to follow in disassembling a piston and connecting rod assembly.
4. List and describe the purpose of the common types of cleaning equipment.
5. Explain how to clean engine parts.

PREPARING TO DISASSEMBLE THE ENGINE

The mechanic should always locate and follow the specific engine removal and disassembly procedures found in a service or shop manual. The procedures presented here are general. They are designed to be used as a guide. The engine may be mounted to a piece of equipment such as a mower or to a vehicle such as a snowmobile or moped. The first job will be to remove the engine.

To prevent a shock or any possibility of fire from a spark, remove the spark plug wire from the spark plug and connect it to the cylinder head. Drain the engine oil from the crankcase and all the fuel out of the fuel tank.

Many small engines have a pulley or blade adapter attached to the crankshaft. These are used to drive the mower blades. A pulley is removed by loosening an allen set screw that holds it in place, Figure 17-1. Vertical crankshaft engines often have a blade adapter attached to the end of the crankshaft. To remove the adapter, thread a bolt into the end of the crankshaft. Attach a puller to the adapter of the crankshaft, as shown in Figure 17-2.

Figure 17-1. Crankshaft pulley is removed by loosening an allen set screw.

Figure 17-2. A puller is used to remove a blade adapter.

When the drive pulley has been removed, clean the output end of the crankshaft. Burrs or rust on the crankshaft may damage the main bearing when the side cover is removed. A narrow piece of abrasive paper is used to polish the crankshaft, as shown in Figure 17-3.

Remove the carburetor and intake manifold assembly. Make sure that the fuel is shut off at the tank if the carburetor and fuel tank are separate. Remove the fuel line and fuel tank. When removing the carburetor, make a sketch of the way the throttle and governor linkage and springs are connected. This will save a lot of time when reassembling. Be careful not to bend or stretch any of the linkage.

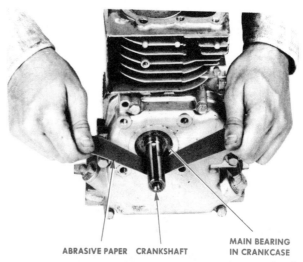

ABRASIVE PAPER   CRANKSHAFT   MAIN BEARING IN CRANKCASE

Figure 17-3. Abrasive paper is used to clean and polish the end of the crankshaft.

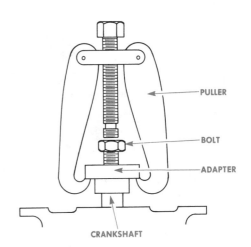

PULLER

BOLT

ADAPTER

CRANKSHAFT

DISASSEMBLING THE TOP OF THE ENGINE

Remove the bolts and screws that attach the cooling shrouds to the engine. Remove the shrouds. Remove the blower housing by removing the bolts that hold it to the engine.

The cylinder head bolts are first loosened and then removed by working from the center of the head outward, Figure 17-4. This prevents the head from being distorted. Never loosen or tighten an aluminum cylinder head when it is hot. This could easily cause the head to be distorted. Notice that some of the head bolts are longer than others. Make a sketch of the location of all the long and short head bolts to help in reassembly. Save the head gasket to match up with the new one.

Before an engine with a cast-iron cylinder can be disassembled further, the cylinder ring ridge must be removed. Most cylinder wear occurs in

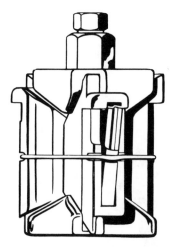

Figure 17-5. The ridge is removed with a ridge reamer. (Ammco Tools, Inc.)

the top inch of ring travel. When this cylinder ridge is large, piston lands may be damaged during removal of pistons from cylinders because the rings sometimes catch on the ridge. Also, the corner of a new top compression ring will strike the rounded lower surface of a large cylinder ridge. This causes a clicking noise when the engine is running and in some cases damages the top ring and bends the second land. The cylinder ridges must be removed with a tool called a *ridge reamer* or *ridge cutter*, Figure 17-5. Aluminum cylinders generally do not have a ridge buildup.

The reamer is driven clockwise with a wrench, Figure 17-6. Follow the instructions supplied by the manufacturer of the ridge cutter. Do not cut down into the ring travel area of the cylinder. It is possible to cut so deeply into the cylinder wall or so far down into ring travel that reboring or replacement of the engine block is necessary. Practically all worn cylinders are out-of-round when cold. But regardless of the unevenness of the wear at the top of ring travel, you should blend the cut made with the ridge reamer so that the area where the machined surface meets the unmachined surface is as smooth as possible. The cuttings from ridge removal should be cleaned away carefully prior to any further disassembly.

Figure 17-4. Remove cylinder head bolts by working from the center outward.

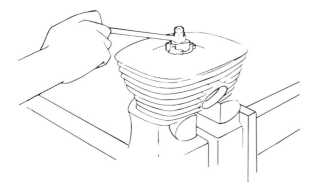

Figure 17-6. The ridge reamer is driven clockwise around the cylinder with a wrench. (Ammco Tools, Inc.)

DISASSEMBLING THE PISTON, CONNECTING ROD AND CRANKSHAFT ASSEMBLY

Remove the flywheel and magneto assembly as described in the unit on ignition service. If the magneto timing is adjustable, be sure to make marks on the magneto to show where it was timed.

The crankshaft end-play must be measured before the side cover is removed. End-play is the side-to-side movement of the crankshaft in the crankcase. There must be enough end-play so the crankshaft can turn freely without binding. Too much end-play may cause excessive crankshaft and main bearing wear. The end-play is adjusted to specifications with gaskets of different thicknesses on the side cover, or with thrust washers between the crankshaft and side cover. The end-play is always measured on disassembly. If the end-play is not correct, it may be changed during reassembly.

Mount a dial indicator to the crankshaft, as shown in Figure 17-7. Set the dial indicator to zero. While observing the dial indicator, move the crankshaft back and forth. The end-play will be shown on the dial indicator. Be sure to record the measurement to check against the specifications.

The side cover is removed by removing the bolts which hold it to the crankcase. Pull it off the crankshaft, as shown in Figure 17-8. If it does not pull off easily, recheck the crankshaft for burrs.

Figure 17-7. A dial indicator is used to measure crankshaft end-play.

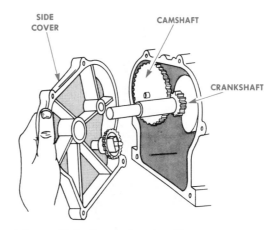

Figure 17-8. Removing the side cover.

With the side cover removed, the crankshaft and camshaft timing gears are visible. The camshaft and crankshaft gears must be in exactly the correct relationship or the engine will not run.

Most engines have timing marks on the two gears to show how they are meshed, Figure 17-9. Wipe off the gears and carefully look for these marks. If you cannot find any marks, use a punch to make your own. After marking, remove the camshaft by pulling it out of its bearing.

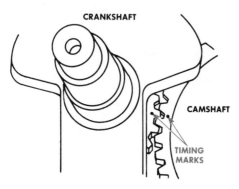

Figure 17-9. Timing marks show proper mesh of camshaft and crankshaft gears.

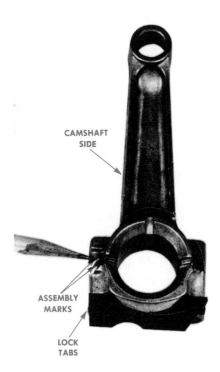

Figure 17-10. Make assembly marks on the rod and cap on the camshaft side.

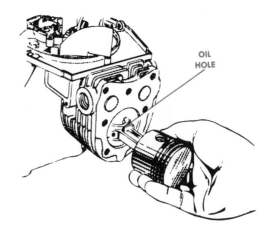

Figure 17-11. The piston and connecting rod assembly is pushed up and out of the cylinder. (Clinton Engines Corp.)

Wipe and inspect the connecting rod bearing caps for a factory marking. Not all engine manufacturers mark the connecting rods. It is good practice to stamp unmarked connecting rods with punches on both cap and rod during removal, Figure 17-10. The markings on the rods should be keyed to the position of the camshaft.

Most connecting rods have a lock tab that must be bent away from the connecting rod bolts or nuts. Remove the connecting rod cap bolts and rod cap from the rod. Push the connecting rod and piston assembly up and out of the top of the cylinder. Care must be taken not to scratch the crankshaft journal. Sometimes it may be necessary to tap the piston assembly up through the cylinder with a hammer handle or block of wood, as shown in Figure 17-11. The crankshaft may then be removed by pulling it out of the main bearing.

The pistons must be disassembled from the connecting rods. The pistons should be marked on their underside to identify the direction in which they fit on the connecting rod. Most engines use full-floating piston pins. Lock rings on both ends of the piston pin hold it in the piston. Remove these retaining rings with needle-nose pliers, Figure 17-12, and push the pin out of the piston and connecting rod. If the pin is tight, the piston may be heated in hot water to loosen it.

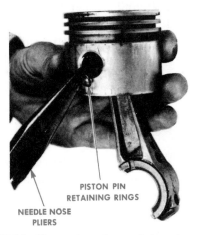

Figure 17-12. Removing pin retaining rings.

PISTON PIN
RETAINING RINGS

NEEDLE NOSE
PLIERS

DISASSEMBLING THE VALVES

The valves are removed by compressing the valve springs with a tool called a *valve spring compressor*, Figure 17-14. The compressor fits either end of the valve spring. Turning the crank on the tool compresses the valve spring, Figure 17-15. This takes the tension off the valve retainer. With the spring compressed, the retainer may be removed. Pin-style retainers are pulled out with needle-nose pliers. Half-circle retainers are removed by pushing them off the valve steam with a screwdriver. Engines that combine the retaining washer and retainer are disassembled by compressing the spring and leaving the retainer loose on the valve stem. The washer has a large hole in the side that allows it to be pushed to the side and

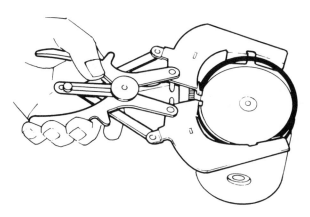

Figure 17-13. Piston rings are removed with a piston ring expander. (Ammco Tools, Inc.)

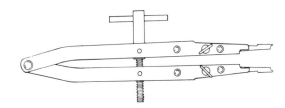

Figure 17-14. Valves are removed with a valve spring compressor. (Ammco Tools, Inc.)

The piston rings must be removed from the piston before cleaning. Piston rings are removed with a piston ring expander tool. The tool fits around the piston. It engages the two ends of the piston ring. When the handle is squeezed, the piston ring is expanded enough to be removed from the piston, Figure 17-13. Do not attempt to remove a piston ring without this type of tool. The ends of the piston ring could scratch and damage the piston. Inspect and mark the rings to indicate the groove in which they fit and which side of the ring faces up. This will help in deciding how the new rings fit.

CRANK

Figure 17-15. The valve spring tool compresses the valve spring.

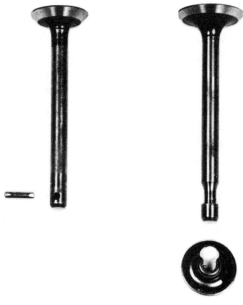

Figure 17-16. Pin- and washer-type valve spring retainers.

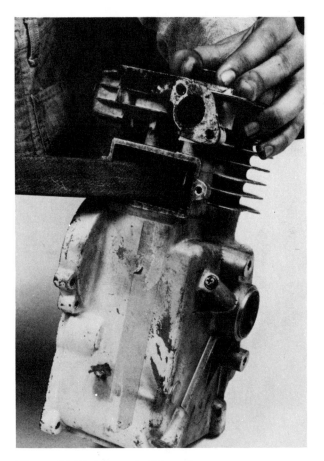

Figure 17-17. The end of the valve must be dressed off with a file before it will come out of the guide.

removed. The pin and washer styles of retainers are shown in Figure 17-16.

The valve may have a burr on the end. The burr must be filed off, Figure 17-17. The valve may then be pulled up out of the valve guide.

CLEANING EQUIPMENT USED FOR ENGINE SERVICE

Small-engine parts are best cleaned with cleaning solvent. Cleaning solvent thins and washes away grease, oil and sludge. A cleaner, or parts washer, Figure 17-18, is used to wash parts in solvent. An electric motor in the cleaner pumps solvent through a hose to flush off parts. The solvent is also circulated through a filter to keep it as clean as possible.

A cold tank cleaner, Figure 17-19, uses a cleaning solution in which the components are soaked for a period of time. Since the cleaning solution is used without heat, the cleaner is referred to as a *cold tank*. Cold-tank cleaning solutions are made to clean nonferrous metal parts such as aluminum and brass. Cold tanks often are used to clean carburetors or aluminum engine parts. The solution in the tank is strong enough to remove carbon and paint. You must wear a face shield, apron

Figure 17-18. Small parts are cleaned in a solvent parts washer. (Gray Mills Co.)

Figure 17-19. Aluminum parts are cleaned in a cold tank. (Gray Mills Co.)

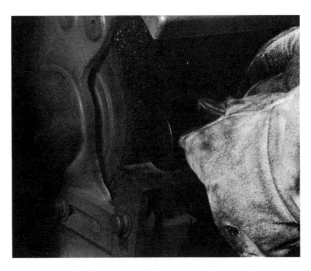

Figure 17-20. Valves may be cleaned on a wire wheel.

and rubber gloves when putting parts into or taking them out of the tank.

CLEANING ENGINE PARTS

The cylinder block is the largest engine component to be cleaned. The block contains many cooling fins and oil passages. Each of these passages must be cleaned thoroughly. A scale builds up on the cooling fins that prevents good heat flow. This scale must be removed so that the rebuilt engine will cool properly. Any foreign material must be removed from the oil passages or it can make its way into the new bearings that will be installed.

The cylinder block may be cleaned in a solvent tank of a cold tank. After the block is cleaned, all deposits must be flushed from the block with water. Oil passageways and holes must be cleaned out with a brush and blown out with air pressure. When the block is dry, oil all the parts that can rust, such as the cylinder liner and valve seats.

Cylinder heads are cleaned in much the same way as the cylinder block. Carbon not removed from a cast-iron combustion chamber by the tanking may be cleaned with a wire brush in a drill motor. Aluminum cylinder heads should not be cleaned with a wire brush since this may scratch the combustion chamber. If a bead blaster is available, it may be used to clean the carbon from the combustion chamber of an aluminum cylinder head. After they are cleaned, all the machined surfaces on the cast-iron cylinder heads must be oiled to prevent rusting. The cylinder head should be covered and stored until time for measurement.

Carbon deposits are removed from the valve by holding it against a wire wheel brush mounted on a grinder, Figure 17-20. The head, face and stem are cleaned thoroughly, particularly the area from face to stem. Carbon can make a valve overheat because it will not allow the heat to flow out.

You must always wear a face shield when using a wire wheel. Also take care not to catch the valve between the wire wheel and the tool rest on the grinder. This could cause the valve to be pulled out of the operator's hand and thrown.

The pistons are made of aluminum. They must be cleaned carefully to prevent scratching the aluminum surface. If there is a buildup of carbon

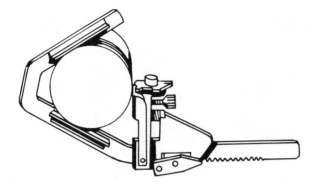

Figure 17-21. Carbon in the ring groove may be removed with a groove cleaner. (Ammco Tools, Inc.)

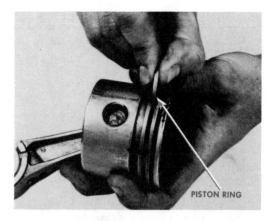

PISTON RING

Figure 17-22. A sharpened piston ring may be used to clean a ring groove.

on the head of the piston, it must be removed by soaking in a cold tank. Do not wire-brush the top of a piston because this will scratch the surface.

A carbon buildup in the ring grooves may be removed with a ring-groove cleaner, Figure 17-21. This tool fits into the ring groove. A scraper is rotated around in the ring groove to remove the carbon. Since there are different sizes of ring grooves, there are different sizes of scrapers on the tool. Care must be taken not to cut into the aluminum on the ring groove.

If a ring-groove cleaner is not available, an old piston ring may be used as a cleaner. The piston ring is broken in two. The end is sharpened and used to scrape out the ring groove, as shown in Figure 17-22.

When the carbon has been removed from the head and ring groove, the piston may be soaked in a cold tank. After several hours of soaking, the piston is removed from the cold tank and flushed with hot water.

The crankshaft may be washed with solvent in a parts washer. A parts-washing brush may be used to loosen deposits. After it is cleaned, the crankshaft must be thoroughly dried. Any oil passages in the crankshaft must be blown clean with compressed air. Dirt or other foreign material left in these passages can ruin new bearings.

When the crankshaft is dry, all the bearing journals must be coated with oil to prevent rusting. If the crankshaft will be stored for a period of time, wrap oiled rags around the bearing journals for protection.

The remaining engine parts should be grouped into those made for ferrous metal and those made from nonferrous metal. Nonferrous parts such as aluminum covers and connecting rods may be cleaned in a cold tank. The ferrous parts such as piston pins, valve lifters, valve springs and washers, as well as the bolts and nuts, also are washed in the solvent parts washer. These parts also must be oiled to prevent rusting.

NEW TERMS

cold tank cleaner: A tank with a cold solution for cleaning nonferrous metal parts such as aluminum.

crankshaft end-play: Movement of the crankshaft measured prior to engine disassembly.

piston ring tool: Tool used to expand piston rings for removal.

solvent cleaner: A cleaning tank in which cleaning solvent is used to wash off oil and grease from small engine parts.

valve spring compressor: Tool used to compress a valve spring so the retainer may be removed.

SELF CHECK

1. Why must the cylinder ring ridge be removed before the piston is removed?
2. Why is crankshaft end-play measured before the engine is disassembled?

3. What should the mechanic do if no timing marks are visible on the crankshaft and camshaft?
4. Describe how to mark connecting rod caps before removal.
5. Describe how to use a valve spring compressor to remove a valve.
6. What type of cleaner is used to clean a cylinder block?
7. In what sequence should the cylinder head bolts be removed?
8. Why should the machined surfaces of a block be oiled after cleaning?
9. Explain how the carbon is removed from piston heads and ring grooves.
10. How are small ferrous metal engine parts cleaned?
11. What parts are cleaned in a solvent parts washer?
12. What parts are cleaned in a cold tank?
13. How is carbon removed from the valves?
14. What safety equipment must be worn when using a cold tank?

DISCUSSION TOPICS AND ACTIVITIES

1. Make and outline the steps to follow in disassembling an engine.
2. Disassemble and clean the parts of a shop engine.

unit 18
valve train service

Valves have to seal well enough to withstand pressures up to 500 pounds per square inch. Under full load, the exhaust valve is exposed to temperatures high enough to cause it to operate red-hot. The temperature of the valve under these conditions may be 1200° F or more. The intake valve is cooled by the incoming mixture. The exhaust valve is subjected to high-temperature exhaust gases passing over it on their way out of the cylinder. It is very difficult to cool the head of the exhaust valve.

In a multi-cylinder engine, one valve may fail and only 1/6th or 1/8th of the power is affected because the bad cylinder may be motorized by the other good cylinders. In a one-cylinder engine, one bad valve can cause a great drop in horsepower or cause the engine to stop entirely. Good valve condition is even more important in one-cylinder engines than it is in multi-cylinder engines. In this unit we will see how to service the valve train.

LET'S FIND OUT: When you finish reading and studying this unit, you should be able to:
1. **Describe how to inspect and measure valve train components for wear.**
2. **Explain how to service valve guides.**
3. **Describe the service procedures used on valve seats.**
4. **Explain how to grind valves.**
5. **List the steps to follow in assembling the valve train and checking for leakage.**

MEASURING AND INSPECTING FOR WEAR

The valve seat and guide must be inspected for wear and damage. Even in normal use, these components absorb a great deal of punishment. High pressures and powerful spring tension pound the red-hot valve head and seat. Hot gases under tremendous pressure swirl past it. Carbon deposits form on the face, preventing the valve from seating properly or cooling efficiently. As a result the valves, particularly the exhaust valves, may become pitted, burned, warped or grooved. They may begin to leak compression and fail to dissipate heat. It is not only the valve face that wears. As shown in Figure 18-1, there is wear at the stem from friction with the guide and end wear from contact with the valve lifter.

The valve seat also wears. Hot gases burn it and carbon particles which retain heat pit it. The valve guide wears against the valve stem. Carbon builds up between stem and guide, which causes the valves to stick. Typical wear in these areas is shown in Figure 18-2.

The amount of wear between the valve stem and guide may be found by measuring the valve

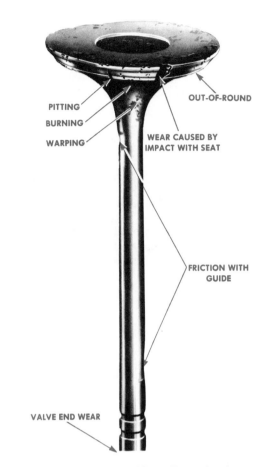

Figure 18-1. Valve wear. (Sioux Tools Inc.)

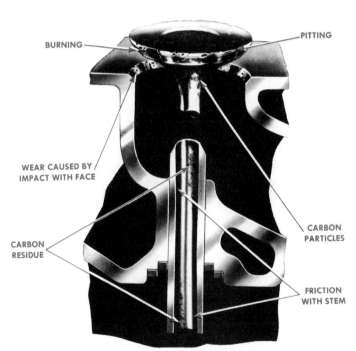

Figure 18-2. Valve seat and guide wear. (Sioux Tools Inc.)

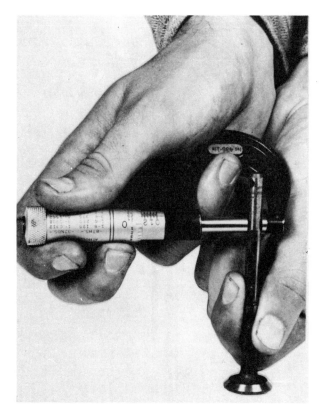

Figure 18-3. The size of the valve stem is determined with a micrometer.

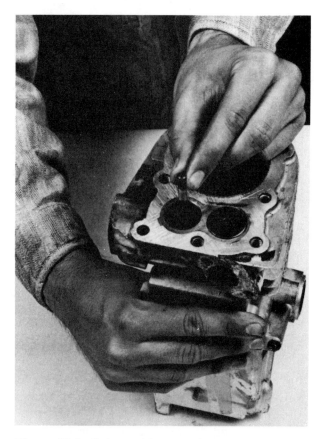

Figure 18-4. Valve guide wear is measured with a small-hole gage.

stem with a micrometer, Figure 18-3, and the guide with a small-hole gage, Figure 18-4. Typically, a valve guide wears more at the top and bottom than in the middle. The valve stem is measured in three places: top, middle and bottom of the stem wear area, and the micrometer readings are recorded. The small-hole gage is inserted into the valve guide and expanded out in contact with the sides of the guide. The small-hole gage is then withdrawn and measured with the micrometer. The guide is measured at the top, middle and bottom. The difference between the smallest valve stem measurement and the largest valve guide measurement is the amount of clearance.

VALVE GUIDE SERVICE

The clearance between the valve stem and guide must allow free movement of the valve. It must also allow a small amount of oil to work its way

between the stem and guide for lubrication. If there is too much clearance, oil from the crankcase area could work its way up the stem and into the combustion chamber. This is a real problem on the intake valve stem because the entire time the intake valve is open, there is a vacuum in the cylinder. On the exhaust valve, there is a vacuum only during the period of valve overlap.

The valve guide also gets rid of heat. Cooling fins in the cylinder block are located near the valve guide area. Heat is moved out of the valve stem, into the valve guide and cylinder block, and into the cooling fins. Excessive clearance will prevent good heat flow. If the stem-to-guide clearance is too large when compared to manufacturer's specifications, there are several repair procedures that may be used.

Many small engines have replaceable valve guides. When the guides are worn, they are

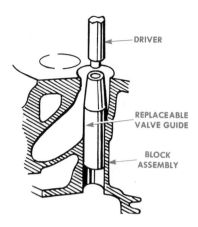

Figure 18-5. Replaceable valve guides are driven in and out of the block assembly with a driver.

Figure 18-6. A reamer is used to machine the new valve guide to size.

replaced with new ones. A worn guide may be removed in several ways. Most often it is driven out. A driver of the correct size is positioned in the guide, Figure 18-5. A hammer is used to drive the guide out through the bottom of the cylinder block.

The new guide is installed by driving it into the cylinder block. The new guide must be driven in carefully to the correct depth. It is good practice to measure how far the old guide sticks out of the cylinder block and to use this measurement for the new guide.

The new guide must be finish-reamed to the correct size. This is done by driving a reamer through it, Figure 18-6. The size of the reamer used is determined by the size of the valve stem and the recommended valve guide clearance. If the valve stem is not worn, a standard-size reamer may be used to establish the correct clearance. If the valve stem is worn, the valve guide will need to be reamed undersize for the correct clearance.

Many small engines do not have replaceable valve guides. The guides are part of the aluminum block casting. The valves operate directly in the aluminum block. There are two service prcedures that may be used on these engines. Some engine manufacturers supply new valves that have stem diameters larger than the original valves. After the new valves are purchased, they are measured with a micrometer. A reamer is selected to machine the guide oversize to fit the new oversize valve stem.

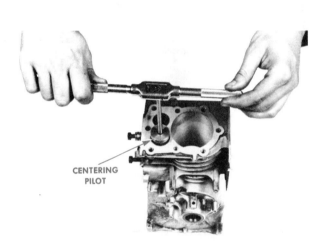

Figure 18-7. Nonreplaceable valve guides may be reamed oversize for thin-wall inserts.

A second repair procedure is to install a thin-wall insert. The thin-wall insert is essentially a new valve guide that has a thin wall. A reamer is used to machine the old valve guide oversize. A centering pilot is used to insure that the reamer is accurate, as shown in Figure 18-7. The reaming is done to a specific depth determined by the length of the new insert. The thin-wall insert is made slightly larger in diameter than the hole machined by the reamer. This provides a press fit to hold the insert in place. The insert is driven into the guide with a driver and hammer, as shown in Figure

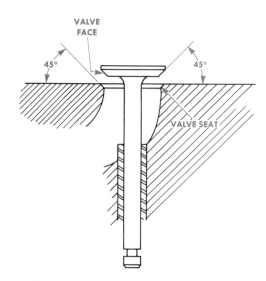

Figure 18-9. The valve seat matches the valve face. (Clinton Engines Corp.)

Figure 18-8. The thin-wall insert is driven into the machined guide.

18-8. It is sometimes helpful to warm up the block under hot water. This expands the block and makes it easier to install the insert. The installed insert is finished to size with a reamer.

VALVE SEAT SERVICE

The valve seat is a precision-ground area at the entrance of the valve port. It may be a part of the cylinder block or a separate unit installed in the block with a press fit. If the cylinder block is made of aluminum, the seats must be made of cast iron or steel. Like valves, the seats may be of a hardened steel alloy called *stellite*.

The angle ground on the valve seat matches the angle ground on the valve face, usually 45 degrees, Figure 18-9. On some engines an angle of 30 degrees is used. Some engines use an interference angle, in which there is a one-degree difference between the seat and face angles. The seat may be ground to 46 degrees and the valve to 45 degrees. Or the seat may be ground to 45 degrees and the valve to 46 degrees. This provides a hairline contact between the valve and seat for positive sealing and reduces buildup of carbon on seating surfaces.

The width of the seat also is important for good sealing, Figure 18-10. If the seat is too wide, there

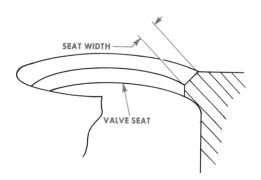

Figure 18-10. Valve seat width is important to good seating. (Briggs & Stratton Corp.)

Figure 18-11. A pilot is installed in the valve guide.

Figure 18-12. The seat grinding stone is dressed on a truing unit.

is a greater chance of a carbon buildup preventing good seating. A wide seat also spreads the valve spring tension over a larger area and reduces heat movement away from the valve head and into the cooling fins near the valve seat. Valve seats may be reconditioned by grinding or cutting. Grinding is the most accurate method but requires the most expensive equipment.

The valve seats may be reconditioned by a grinding stone mounted in a holder and driven by a hand driver. Always wear eye protection when operating this equipment. A pilot shaft of the correct diameter is inserted in the valve guide of the seat to be ground, Figure 18-11. The pilot is used to guide and center the grinding wheel. Since the valve guide is used for centering, all valve guide service or replacement must be performed before seat grinding.

A seat grinding stone is mounted to a driver and dressed to the specified angle for the valve seat in a truing unit, using a diamond cutting tool, Figure 18-12. The dressing tool is adjusted to the proper angle. Put a drop of very light oil on the dressing pilot to prevent sticking. (Do not get oil on the grinding wheel.) Screw the grinding wheel on its holder and place on the pilot. Adjust until the wheel just touches the diamond. Insert the

Figure 18-13. The grinding stone is placed over the pilot.

driver and bring the wheel up to speed. Move the diamond steadily across the wheel, taking light cuts. Hold the driver as straight as possible.

The grinding stone and holder are installed over the pilot, as shown in Figure 18-13. The

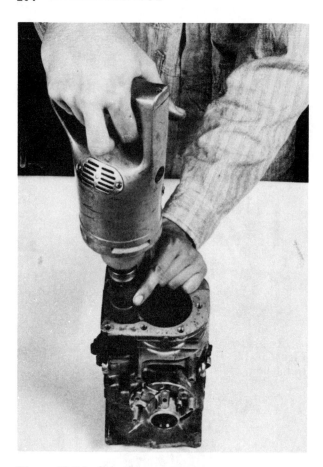

Figure 18-14. Grinding a valve seat.

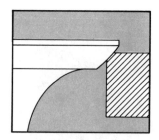

Figure 18-15. Proper valve face-to-seat contact.

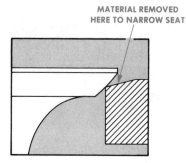

MATERIAL REMOVED
HERE TO NARROW SEAT

Figure 18-16. A valve seat narrowed at the top.

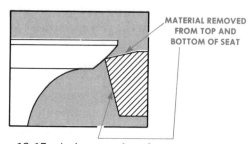

MATERIAL REMOVED
FROM TOP AND
BOTTOM OF SEAT

Figure 18-17. A three-angle valve seat.

driver spindle is installed on the holder as shown in Figure 18-14, and the driver motor started. Only very light pressure is required when grinding. Support the weight and let it run at high speed. When grinding, sway the top of the driver gently from side to side—about 1/4″ off-center to each side. Do not use pressure. Excessive grinding may go below the hardened surface. These hardened seats will dull grinding wheels rapidly. The grinding wheel must be dressed frequently when servicing these hardened seats.

The ground valve seat must provide the proper seat-to-face contact, as shown in Figure 18-15. The valve face should always be larger than the valve seat. The seat should be wide enough to assist the valve in dissipating heat but not wide enough to collect carbon deposits. The engine manufacturer will provide specifications on valve seat width. If the reconditioned seat is too wide, it must be narrowed. Narrowing is done by grinding material off the top or bottom of the seat with a special narrowing grinding stone. The valve seat shown in Figure 18-16 has been narrowed at the top for proper seat contact using a 15-degree stone. When material is removed from both the top and bottom, it is described as a three-angle seat. A three-angle valve seat is shown in Figure 18-17.

Valve seats may be reconditioned with a hardened steel cutting tool such as the one shown in Figure 18-18. The cutter has several sharpened

Figure 18-18. Valve seats may be reconditioned with a cutter. (Clinton Engines Corp.)

steel blades that remove a small amount of metal from the seat. The cutter usually has one set of blades for 45-degree seats and another for 30-degree seats. A variety of cutters is necessary for seats of different diameters.

A pilot is installed in the valve guide. The cutter is installed over the pilot. The cutter is driven by hand with a T-handle wrench, as shown in Figure 18-19.

When reworking valve seats, it is necessary to cut away all of the oxidized metal until new and solid metal is exposed. A good seat normally will have a brighter appearance than surrounding metal. Using a cutter, apply steady pressure directly downward to minimize the possibility of not having the seat true to the guide. Excess pressure can cause the cutter to chatter and make the seat unsuitable for use. It is difficult then to remove the irregularity caused by chatter and make a seat suitable for sealing to the valve face.

Another much less precise method is used in some shops to recondition valve seats. An abrasive compound is spread on the valve face and valve seat. The valve is placed down on the seat and rotated back and forth with a small suction cup unit with a handle. The abrasives in the compound remove metal from the valve and seat. The procedure is called *lapping*. Lapping matches the contour of the valve face with that of the valve seat. The problem is that as the valve heats up in actual operation, it changes its shape and the contours no longer line up. For this reason, lapping is not a good way to recondition a seat. It should be used only when other types of equipment are not available.

Replacing Valve Seats

If a valve seat is very badly pitted, the grinding operation may not be enough for reconditioning. In this case, the valve seat may have to be replaced. If the cylinder block has a replaceable valve seat, a new one can be installed. If the block does not have a replaceable valve seat, the block may be machined to accept one.

Replaceable valve seats are removed with a valve seat puller. The puller rests on top of the cylinder block. A puller nut fits under the valve seat insert, as shown in Figure 18-20. The bolt of the puller drives the nut up and pulls the valve seat insert out of the block.

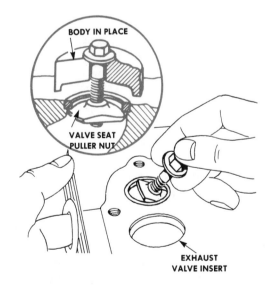

Figure 18-20. Valve seat inserts are removed with a puller. (Briggs & Stratton Corp.)

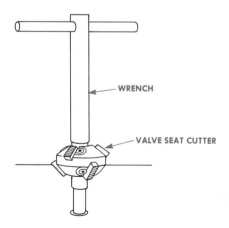

Figure 18-19. Cutting a valve seat is done by hand. (Briggs & Stratton Corp.)

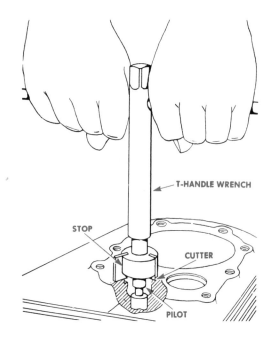

Figure 18-21. Machining the cylinder block for a new valve seat insert. (Briggs & Stratton Corp.)

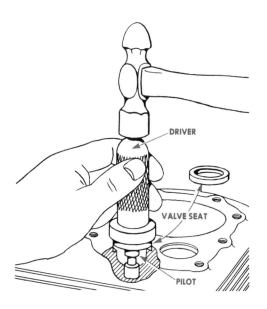

Figure 18-22. Installing a new valve seat. (Briggs & Stratton Corp.)

A block without a replaceable valve seat may be machined with a cutter. A pilot is installed in the valve guide. A cutter, similar to that used for valve seats, is driven by hand to machine the area for a new valve seat, Figure 18-21.

A new insert is installed by driving it with a hammer and special driver, Figure 18-22. The driver has a pilot on the end to insure that the seat is centered in relation to the valve guide. The new valve seat is driven into the machined hole until it bottoms. The new seat will require a slight grinding or cutting to make sure it is in alignment with the valve guide.

VALVE SERVICE

Valves are made from very high-quality steels because they get very hot during combustion. The intake valve usually is larger in diameter than the exhaust valve because it must control the slow-moving, low-pressure intake mixture. Exhaust valves may be smaller because the exhaust gases leave the cylinder under higher pressures. Since the exhaust valve gets even hotter than the intake valve, it is made from even higher-quality steel. Stainless and stellite steel alloys often are used.

The valve is serviced on a machine called a *valve grinder*, Figure 18-23. The valve grinder has two grinding wheels, one for grinding the valve

Figure 18-24. The tip is resurfaced by passing it over the side of the wheel.

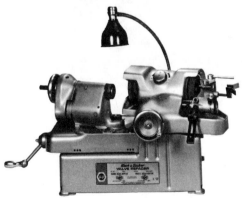

Figure 18-23. A complete valve and seat grinding set. (Black & Decker)

face and the other for valve stems, rocker arms and other valve train components. Always wear eye protection when using this equipment.

The first operation in valve service is to regrind and chamfer the valve stem tip. This operation, called *stemming the valve*, is necessary to insure proper centering of the valve. The valve is clamped on its stem in a V-bracket. It is advanced toward the side of the grinding wheel with a micrometer feed. Coolant is pumped over the wheel and valve during grinding. When the stem end contacts the wheel, the valve is moved back and forth across the wheel side, Figure 18-24. Just enough material is removed to resurface the tip. After grinding, the tip is chamfered, or cut to a slight angle. The valve is positioned in a fixture and advanced toward the wheel. It is rotated by hand to grind a slight chamfer into the tip.

The valve face grinding operation begins by "dressing" or truing the valve face grinding wheel. The dressing is accomplished with a diamond mounted to a tool post. The wheel should be

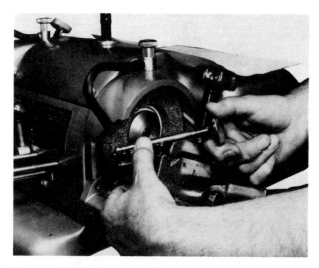

Figure 18-25. The valve is inserted into the chuck so the rollers engage the stem in the unworn area.

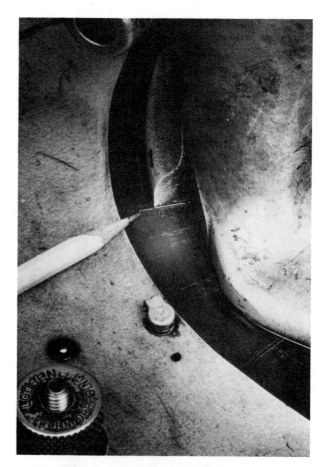

Figure 18-26. The chuck is adjusted to the correct angle with a hold-down nut.

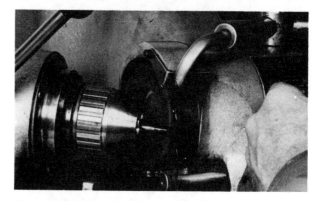

Figure 18-27. The valve is ground by moving it slowly back and forth across the wheel.

dressed each time the grinding head is installed. Unless the valves are very hard, one dressing is enough for a complete set of valves.

After dressing, the valve may be installed in the chuck. The chuck is designed to grip the valve in the unworn part of the stem. The valve is centered in the chuck off the valve tip. Open the chuck sleeve and insert the valve so that rollers will engage the stem just above the worn area, Figure 18-25.

The valve chuck must be adjusted to the correct angle. Specifications may call for the valves to be ground to 45 or 30 degrees. An interference angle may be required. Graduations on the chuck carriage allow the chuck to be indexed to the correct angle, Figure 18-26. A hold-down nut is loosened, and the chuck moved into position. The hold-down nut is then retightened.

Begin grinding at the left side of the wheel, moving the valve slowly and steadily, right and left across the wheel, Figure 18-27. Do not allow the valve at any time to pass beyond either edge of the grinding wheel while grinding. Take light cuts by feeding the wheel up to the valve about .001 inch to .002 inch at a time. Remove just enough material to make a clean, smooth face. When the valve face is trued, advance it to the right until the tip edge of the valve is flush with the right edge of the grinding wheel. Pause a second, then back the

grinding wheel away from the valve, *not the valve away from the wheel.*

After grinding, inspect each valve. The margin on the valve is reduced when the valves are

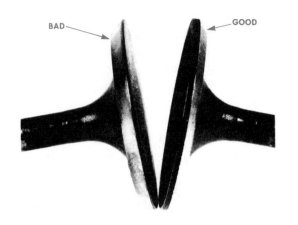

Figure 18-28. A valve with too thin a margin after grinding must be replaced.

ground. Badly pitted, burned or warped valves cannot be refaced without removing too much of the margin, Figure 18-28. Loss of the margin would provide a sharp edge on the head of the valve that would not be able to take high temperatures. If margins are not to specified thickness after grinding, the valve must be replaced.

VALVE SPRING SERVICE

The valve spring is tested for squareness by placing it alongside a combination square, as shown in Figure 18-29. The spring is rotated into several positions. If it is out of square there will be a space between the spring and square at the top. An out-of-square spring must be replaced.

VALVE LIFTER SERVICE

The valve lifter is the part of the valve train that rides on the camshaft. The lifter must be inspected carefully for wear. The most likely surface to wear is the bottom of the lifter where it contacts the camshaft. Cupping and pitting are common types of wear in this area. If there is any evidence of surface deterioration, the lifter should be replaced. The lifters also should be checked for fit in the lifter bores of the block. Excessive clearance here will require that new and perhaps oversize lifters be installed.

SETTING VALVE CLEARANCE

After the valve seat and valve have been reconditioned, the valve lash must be adjusted. This job usually is done when the engine is being reassembled and after the camshaft and lifters have been installed. Valve lash, or valve clearance, is the space allowed for expansion in a valve train. If the valve lash is too wide, there will be engine noise and wear on the camshaft and lifter contact faces. Eventually, valve timing is affected and the engine performs poorly.

If the valve lash is too small, a valve may be prevented from closing tightly on its seat and sealing the combustion chamber. The immediate result is poor engine performance. But also, a valve held off its seat gets very hot because it is not able to move heat away from the valve head into the valve seat in the cylinder block. The result is a melted valve head, called a *burned valve*. A

Figure 18-29. The valve springs are checked for squareness with a square.

burned valve cannot seal, so the engine performs poorly.

Valve lash is adjusted in most small engines by grinding the end of the valve stem. Rotate the

Figure 18-30. Valve lash is checked with a feeler gage.

crankshaft until the camshaft lobes are pointed away from the lifters. This insures that the valve is in the closed position. Then insert the valve into its guide. With your hand push down the valve while measuring the space between the lifter and valve stem with a feeler gage, as shown in Figure 18-30. After the valve face and seat are ground, this distance normally will be closer than specifications require. Remove the valve from the engine and place it on a valve grinder, as shown in Figure 18-31. A small amount of material is removed from the end of the stem. The valve lash is checked again and more material removed if necessary.

VALVE REASSEMBLY AND TESTING

The valve and spring assembly may now be reinstalled into the reconditioned cylinder. Reassembly is done with the same valve spring com-

Figure 18-31. Valve lash is adjusted by grinding the end of the valve stem.

Figure 18-32. Installing the valve assembly.

Figure 18-33. Valve face-to-seat seal may be checked with solvent.

pressor used for disassembly. Remember to reinstall the valves in the same positions from which they were removed. Lubricate the valve stem with engine oil and insert the valve into the valve guide. Install the valve spring over the valve stem and compress it with the compressing tool. Reinstall the retainer washer. Position the keepers on the stem. The compressing tool is slowly released, as shown in Figure 18-32.

When the valve assembly is reassembled, the seal between the valve face and seat may be tested. Place the cylinder block on a level bench. Fill the area around the valves with clean solvent, Figure 18-33. The solvent should not leak out between the valve and seat. A leaking valve and seat will have a steady leak around the head of the valve. In this case, the valve and seat will require regrinding or lapping.

REED VALVE SERVICE

As described earlier, two-stroke-cycle engines use a reed or rotary valve to control air-fuel mixture flow. The reed valve and reed plate assembly should be inspected when the engine is disassembled. Check the reed on the reed plate for cracks or distortion. The reeds should lie flat on

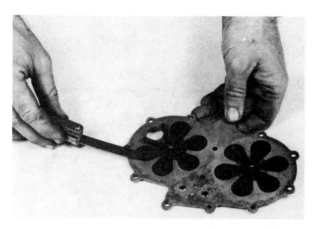

Figure 18-34. Reed valve opening is checked with a feeler gage.

the reed plate and should not bend away from the plate excessively. This distance may be checked with a feeler gage, Figure 18-34, and compared to manufacturer's specifications.

NEW TERMS

margin: The outside part of the valve face that gets thinner as it is ground.

narrowing: Removing part of the valve seat to make it narrow for better valve seating.

thin-wall guide: A thin bushing, or metal sleeve, used to recondition a worn valve guide.

valve guide clearance: The space between the valve guide and valve stem.

valve grinding: Machining the valve face by grinding.

valve lash adjustment: The setting of the valve clearance or lash to the correct specifications.

valve seat cutting: Reconditioning a valve seat by cutting with a hardened steel cutting tool.

valve seat grinding: Reconditioning a valve seat by grinding with a grinding stone.

valve seat lapping: Using an abrasive compound to remove metal from a valve seat and face.

valve spring tension: The tension of a valve spring measured when the spring is compressed.

SELF CHECK

1. How are valve guides measured for wear?
2. Explain how to remove and replace valve guides.
3. Describe how to dress a seat grinding stone.
4. Why must valve guides be serviced before valve seats?
5. List the steps to follow in grinding a valve seat.
6. Describe how to recondition a valve stem.
7. List the steps to follow in grinding the face of a valve.
8. Why is a valve with a thin margin replaced?
9. How is a valve spring tested for squareness?
10. Describe how to test a valve face and seat for leaks.

DISCUSSION TOPICS AND ACTIVITIES

1. Disassemble and reassemble a small engine valve assembly.
2. Use the equipment available in your shop to recondition a valve and seat.

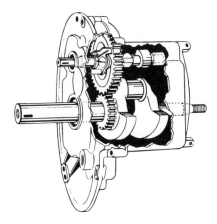

unit 19

crankshaft, camshaft, rod and bearing service

After you disassemble and clean each of the engine parts, you must inspect them for wear. Worn components must be replaced or reconditioned. In this unit we will see how to inspect and recondition the crankshaft, camshaft, connecting rod and main bearings.

LET'S FIND OUT: **When you finish reading and studying this unit, you should be able to:**
1. **Describe how to measure a crankshaft for wear.**
2. **Explain how to inspect a camshaft for wear.**
3. **Describe how to measure a connecting rod for wear.**
4. **Explain how to find the amount of wear in a main bearing.**
5. **Describe how to recondition main bearings.**

CRANKSHAFT INSPECTION AND MEASUREMENT

The crankshaft must be thoroughly inspected and measured. Visually inspect the crankshaft for evidence of damage as shown in Figure 19-1. Crankshaft journals have a tendency to wear out-of-round as well as to reduce in size. Rod journals usually show the most wear on the underside of the "throw," but main journals have no particular area of wear. Carefully inspect each journal for scoring. Even the smallest score marks will require that the crankshaft be reground.

If the journal surfaces are smooth, the next step is to carefully measure with an outside micrometer. Measure the diameters of the journals,

Figure 19-1. The crankshaft is measured with an outside micrometer.

213

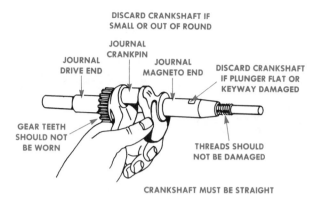

DISCARD CRANKSHAFT IF
SMALL OR OUT OF ROUND

JOURNAL
CRANKPIN

JOURNAL
DRIVE END

JOURNAL
MAGNETO END

DISCARD CRANKSHAFT
IF PLUNGER FLAT OR
KEYWAY DAMAGED

GEAR TEETH
SHOULD NOT
BE WORN

THREADS SHOULD
NOT BE DAMAGED

CRANKSHAFT MUST BE STRAIGHT

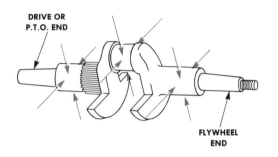

DRIVE OR
P.T.O. END

FLYWHEEL
END

Figure 19-2. The crankshaft is measured in several places. (Briggs & Stratton Corp.)

about one-quarter inch in from each end or enough to clear the fillet radius, and around the journal in several places, as shown in Figure 19-2, to get the smallest and largest readings.

When a journal is larger in one direction than another, it is tapered. Record each measurement, indicating the minimum and maximum sizes. It is best to do this in chart form so that maximum out-of-roundness and taper can be determined. Start with the front or power takeoff (PTO) journal. If any journal measures .001 inch out-of-round, or has more than .001 inch taper, or if any journal is rough or ridged, the crankshaft should be reconditioned or replaced.

Crankshaft Reconditioning

A perfectly round, smooth and straight crankshaft is best for proper bearing performance. To regrind crankshaft journals to the necessary degree of accuracy requires special precision machinery and an experienced machinist.

Crankshafts are reconditioned in a precision grinder. Each of the connecting rod and main bearing journals is ground to a standard undersize. The bearings used with the reconditioned crankshaft must be matched to the undersize so that excessive oil clearance will not result. Too much clearance between bearing and shaft will result in knocking and rapid wear of shaft and bearing.

CAMSHAFT INSPECTION AND MEASUREMENT

The camshaft lobes have a very difficult job to do. They must push the lifters up against strong spring pressure. They absorb a constant pounding as the valve lash changes. These forces cause wear on the cam lobe surfaces. Any evidence of pits or scuffing will require that the camshaft be replaced.

An outside micrometer may be used to compare the lobe heights. Place an outside micrometer in position to measure from the heel to the top of the lobe. Figure 19-3. Record the measure-

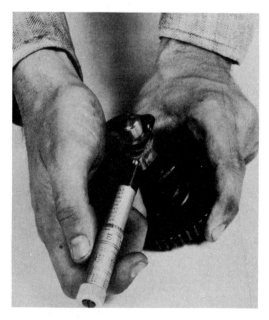

Figure 19-3. An outside micrometer may be used to compare lobe heights.

ment. Make a similar measurement on all the intake and exhaust lobes. This measurement is compared to specifications to determine the amount of wear.

The camshaft bearing journals should be inspected. Any evidence of scuffing or scratching will require reconditioning or replacement. Use an outside micrometer to measure the diameter of each of the journals. These measurements are compared with a measurement of the camshaft bearings in the block and side cover. Measure the bearings with a small telescoping gage. The difference between the camshaft journal and bearing diameters is the amount of clearance. Compare this clearance to specifications.

CONNECTING ROD AND PISTON PIN INSPECTION

When the connecting rods have been cleaned, inspect them visually for any evidence of damage. Rods with excessive scoring, such as that shown in Figure 19-4, should be replaced. If a separate bearing is used, remove it, replace the rod cap in the rod, and tighten to specifications. Use a connecting rod vise, Figure 19-5, to insure the correct alignment of the rod to cap. Then inspect the

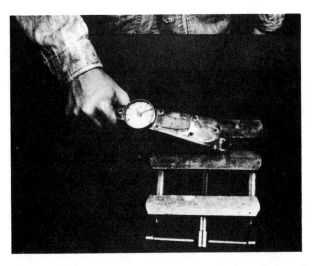

Figure 19-5. Connecting rod caps are installed and tightened in a connecting rod vise.

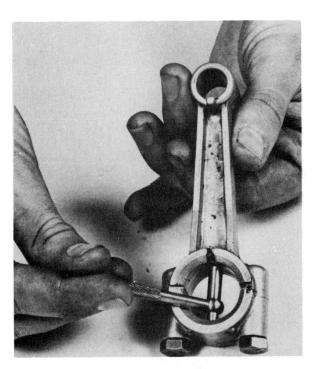

Figure 19-6. The connecting rod saddle is measured for out-of-round with a telescoping gage.

Figure 19-4. Rods with excessive scoring must be replaced.

crankshaft end of the connecting rod (called the big-end bore or saddle bore) for out-of-roundness. This may be determined by taking several measurements around the bore with a telescoping gage, Figure 19-6.

A condition of out-of-round at the saddle bore usually is described as *stretch*. During the power stroke, the top of the saddle bore is placed under a severe load. The bottom of the saddle bore is in a no-load area. This uneven load application results in a saddle bore that eventually stretches out-of-round. A connecting rod with a stretched saddle bore will have a very short service life.

Measurement of the condition of the small end of the connecting rod also is very important. At this end, the concern is how well the piston pin fits the connecting rod. The piston pin, often called a *wrist pin,* connects the piston to the connecting rod by fitting through holes in the piston and connecting rod bosses. The pin allows the rod to swivel back and forth with respect to the piston. The full force of the combustion pressures is transferred from the piston to the connecting rod through the small piston pin. The pin is made from a high-quality steel, usually in tubular form for lighter weight.

The pin is slightly smaller than the hole either in the piston or in the connecting rod. It is free to "float" to rotate in both parts. The pin is held with two internal lock rings, one on each side of the piston. The advantage of this design is that the pin

could stick in either the piston or rod without the assembly locking up. The "fit" or clearance between the pin and rod and piston is one of the closest and most precise in the engine.

Check full floating pins for clearance in both the piston and connecting rod. The fit is checked by inserting the pin into the connecting rod and piston. Try to rock the pin up and down, Figure 19-7. Any noticeable movement means excessive wear in the connecting rod. It is most likely that wear will occur in the rod since there is less bearing surface area here than in the piston. This wear usually is corrected by replacing the connecting rod.

CONNECTING ROD BEARING INSPECTION

Many small engines do not have a separate connecting bearing. The aluminum connecting rod saddle bore is used as the bearing surface. Many two-stroke engines use needle bearings in the saddle bore, Figure 19-8. The needle bearings may be free to touch each other or they may be held apart with a cage. Inspect each of the needle bearings for signs of scoring. Bearing clearance is determined by measuring each of the needle bearings with a micrometer. This measurement is compared to specifications to determine whether the connecting rod clearance is excessive. Worn needle bearings must be replaced.

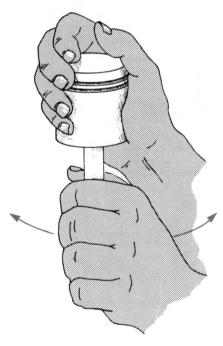

Figure 19-7. Checking piston pin fit.

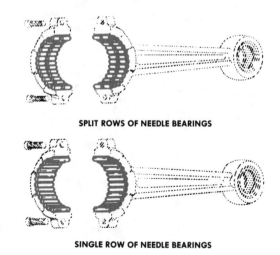

SPLIT ROWS OF NEEDLE BEARINGS

SINGLE ROW OF NEEDLE BEARINGS

Figure 19-8. Needle bearings sometimes are used in two-stroke connecting rods. (Tecumseh Products Co.)

Figure 19-9. An insert bearing is used in some connecting rods.

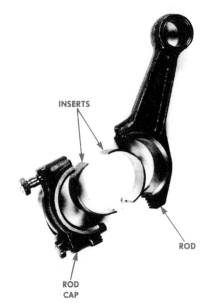

Some small engines use a precision insert bearing between the crankshaft and connecting rod bore. The insert bearings are made in two pieces so they may be assembled around the crankshaft, Figure 19-9. The insert bearing consists of a steel backing with a thin layer of softer bearing material attached to the backing, Figure 19-10. Most insert bearings have what is called "spread." They are slightly larger than the housing in which they fit. The spread allows them to snap into place. Besides spread, bearings are provided with some other means of locking in the housing. Many

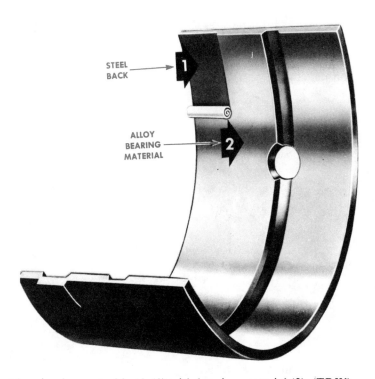

Figure 19-10. An insert bearing has a steel back (1) with bearing material (2). (TRW)

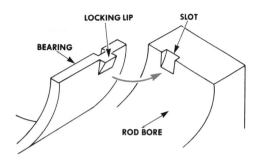

Figure 19-11. An insert bearing must be locked into the rod bore.

bearings have a locking lip that fits into a slot in the bore, Figure 19-11. Others are held with a dowel pin in the housing that fits into a dowel hole in the bearing.

Always replace insert bearings during an engine overhaul. The bearings come in standard sizes to be used in engines with a standard-size crankshaft. Thicker inserts are used to reduce excessive clearance or with crankshafts that have been ground undersize. These bearings are called *undersize bearings*.

MAIN BEARING INSPECTION

In many small engines, the crankshaft operates directly in the aluminum block and side cover. Inspect each bearing for scoring. If the bearing surface is scored or damaged, it must be reconditioned.

Figure 19-12. Main bearings are measured with a telescoping gage.

If the bearing passes a visual test, it may be measured. A small telescoping gage is used along with a micrometer, Figure 19-12. Measure the bearing in several places to determine whether it is out-of-round or worn undersize compared to specifications. If so, the bearing must be reconditioned.

Main Bearing Replacement

Some engines use tapered roller, ball or needle bearings to support the crankshaft, Figure 19-13. The more expensive small engines typically use a ball bearing on the power takeoff end of the crankshaft. This bearing is subjected to severe loads when the engine drives mowing equipment. Two-stroke engines often use needle bearings on both mains. Larger stationary engines that turn relatively slowly often use a tapered roller bearing on the mains.

The ball and roller bearings fit on the crankshaft with a press fit. They should be left in place when cleaning the crankshaft. After cleaning, the

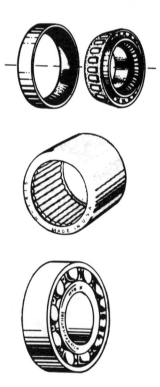

Figure 19-13. Types of main bearings.

bearings should be relubricated and rotated to check for damage. The tapered roller races mounted in the block, and the bearing plate, should be washed in solvent, dried and inspected for wear or roughness. If the race needs to be replaced on either the bearing plate or block, pull or drive the race out, using care not to damage the block or side cover. Press a new race back into the block or side cover.

If the roller or ball bearings do not pass visual inspection or are rough after cleaning and lubrication, they can be removed from the crankshaft with a bearing spreader. This spreader breaks the bearing free from the crankshaft by tightening down the two burrs which will start the bearing race moving. Then the bearing can be pulled with a puller hooked to the spreader, as shown in Figure 19-14. Replace the bearings by pressing

Figure 19-14. A bearing spreader and puller are used to remove a roller bearing.

Figure 19-15. A new bearing is installed in a press.

them on the crankshaft carefully so as not to damage them. Support the crankshaft so that the crankpin is not distorted by the pressure needed to replace roller or ball bearings, Figure 19-15.

Reconditioning Main Bearings

A main bearing that is part of the cylinder block or side cover is reconditioned by reaming and installing a bushing. The reaming operation must be done so that both main bearings are in perfect alignment. If the reconditioned main bearings are not in correct alignment, the crankshaft will not rotate freely. Each engine manufacturer supplies a set of reamers with pilots that insure correct alignment. The proper tools always must be used for a main bearing job.

Install the side cover and torque the proper amount before reaming. Install a pilot-guide bushing in one of the main bearings to center and guide the reamer in the other bearings. Install the correct size reamer on a pilot. The pilot fits into the guide bushing in the opposite main bearing and a large guide bushing in the main bearing housing being reamed, as shown in Figure 19-16. Turn the reamer clockwise by hand to machine the bearing housing oversize. The two guide bushings insure that the reaming is done in correct alignment. Reverse procedure with a different set

Figure 19-16. Main bearings are reamed for new bushings.

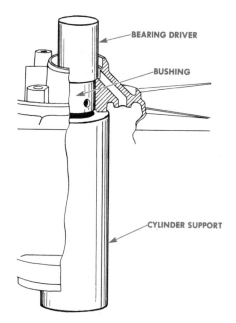

Figure 19-17. Tools used when driving in a main bearing bushing. (Briggs & Stratton)

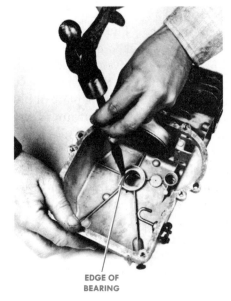

Figure 19-18. A bushing must be staked to prevent its turning.

of guide bushings to ream out the other main bearing.

After reaming, use a bearing driver to drive the bushing into the main bearing housing. Back up the block during driving with a cylinder support, Figure 19-17. Align the bushing with any oil feed notches or holes. With a blunt chisel, drive a part of the bushing into the notch previously made in the cylinder, Figure 19-18. This is called *staking* and is done to prevent the bushing from turning.

Reinstall the guide bushings and attach a finish reamer to the pilot. Ream each main-bearing bushing so that the inside diameter is the correct size according to specifications. Finish reaming often is done with a lubricant such as kerosene on the cutter, to prevent a rough finish.

NEW TERMS

crankshaft out-of-round: A wear condition in which a crankshaft journal is no longer round.
crankshaft taper: A wear condition in which one end of a crankshaft journal is larger than another.

fit: A term used to describe the clearance between parts.
guide bushing: Tool used to guide a reamer when reconditioning main bearings.
lobe height: The distance between the heel and the top of a cam lobe.
multi-piece crankshaft: A crankshaft that may be disassembled to replace a journal.
saddle bore: The large hole in the bottom of the connecting rod that fits around the crankshaft.

SELF CHECK

1. Describe the types of wear common to crankshafts.
2. Describe how to measure a crankshaft for wear.
3. How is a crankshaft reconditioned?
4. Explain how to measure a camshaft for wear.
5. What forces cause the camshaft lobes to wear?
6. How is connecting rod saddle bore out-of-round determined?

7. Describe how to check for proper piston pin clearance.
8. How are main bearings measured for wear?
9. Describe how to remove a ball-type main bearing from a crankshaft.
10. Explain how to recondition a main bearing.

DISCUSSION TOPICS AND ACTIVITIES

1. Using practice parts available in the shop, measure a crankshaft, camshaft and connecting rod for wear.
2. Look up wear tolerances and part specifications for an engine you own.

unit 20

cylinder, piston and ring service

During engine operation, the cylinder walls are subjected to tremendous forces. Piston rings are continually pushing against and scraping them. Combustion forces continually thrust the piston against them. The cylinder walls are dependent upon splash lubrication in areas that are the hottest in all the engine. The piston and piston rings accept the full force of the burning gases during the power stroke. These forces combine to make the cylinder walls, piston and rings the fastest wearing areas of the engine and some of the most important areas of service. In this unit we will see how to determine the amount of wear in these parts and how they are reconditioned.

LET'S FIND OUT: **When you finish reading and studying this unit, you will be able to:**
1. **List the steps to follow to determine cylinder wear.**
2. **Describe how to break the glaze on the cylinder wall.**
3. **Explain how to recondition a cylinder wall by honing and boring.**
4. **List the steps used to determine if a piston is worn excessively.**
5. **Explain how to measure piston ring end gap.**

CYLINDER INSPECTION AND MEASUREMENT

When the block is cleaned, it is ready for inspection. The block must undergo a thorough inspection and measuring to determine what reconditioning will be necessary. Inspect the cylinder for possible scoring from broken piston pin lock rings or piston rings. If no scoring is evident, measure the cylinders to determine their size and the amount of wear.

Wear for a typical cylinder is shown in Figure 20-1. The area of greatest wear is found where the piston rings operate above the upper end of piston skirt travel. This area, often called the *pocket*, receives the least lubrication and is subjected to

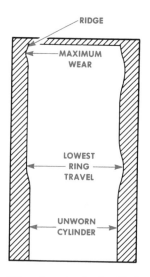

Figure 20-1. Wear in a typical cylinder.

the forces from the piston rings. The area of least wear is below the upper end of piston skirt travel. The area at the very bottom of the cylinder is below ring travel and not subject to much wear. This is called the unworn part of the cylinder.

The difference between the wear at the top of the cylinder and the wear at the bottom is called *taper*. As the crankshaft turns, the pistons constantly are being thrust at the sides of the cylinder. This action also causes the cylinder to become egg-shaped or out-of-round.

Measure the cylinder with a telescoping gage and micrometer, or a special dial indicator called a *cylinder gage*. If a telescoping gage is used, choose one that will expand to the size of the cylinder. Follow the instructions in the unit on using and reading micrometers and telescoping gages.

The cylinder gage, Figure 20-2, has a set of rails that fit against the cylinder wall. A plunger, attached to a dial indicator, pushes out against the

cylinder. The gage assembly is positioned in the bottom of the cylinder. The plunger and dial indicator are set to read zero in the unworn section of the cylinder. The gage may be moved up or around the cylinder. The dial indicator will register variations in the cylinder size.

Regardless of the type of measuring tool used, the same measurements are made. Measure the area of most wear in the direction the crankshaft is installed and in the direction opposite the crankshaft. The difference between these two measurements is the amount of out-of-round. A measurement is made at the bottom of the cylinder bore and compared with one in the area of most wear. The difference between the two measurements is called *taper*. The measurement at the bottom where the cylinder bore is unworn may be compared to specifications to find the size of the cylinder. This information is necessary when ordering piston rings. The areas where measurements are made are shown in Figure 20-3.

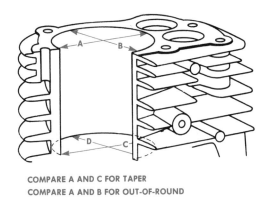

COMPARE A AND C FOR TAPER
COMPARE A AND B FOR OUT-OF-ROUND

Figure 20-3. Where to make cylinder measurements.

If a telescoping gage or cylinder gage is not available, cylinder taper may be found with a feeler gage. This method is not as accurate as precision measuring but will give a general indication of cylinder wear. Use a piston ring from the engine to be measured. Place the piston ring in the cylinder. Push it down into the cylinder about an inch, using the head of the piston. Make sure the ring is square in the cylinder. Measure the space between the two ends of the ring with a feeler

Figure 20-2. A cylinder gage provides a dial reading of cylinder taper.

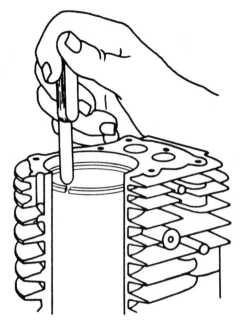

Figure 20-4. Finding taper with a feeler gage and piston ring. (Briggs & Stratton Corp.)

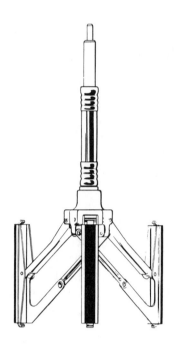

Figure 20-5. Glaze breaker with spring-loaded stones. (Ammco Tools, Inc.)

gage, as shown in Figure 20-4. Move the piston down into the unworn part of the cylinder. Measure the gap here. Subtract the reading at the bottom from the one at the top and divide by 3. This is *cylinder taper.*

GLAZE-BREAKING

The cylinder wear measurements are compared against manufacturer's specifications to find out what reconditioning will be necessary. If taper and out-of-round are within limits, all that is necessary is to deglaze the cylinder walls. The movement of the piston rings up and down in the cylinder polishes the cylinder surface. This polish or glaze must be removed so that new rings will wear in or "seat" quickly. Also, oil will cling more readily to a deglazed surface and prevent ring or piston scuffing.

Some piston rings have a coating of molybdenum sprayed on the ring surface. The coating helps prevent scuffing and wear. These rings are called "moly" rings. In some cases the piston ring manufacturer may specify that the cylinder not be deglazed for "moly" rings as the rough cylinder finish may destroy the coating. The mechanic

must always check the installation instructions carefully when using these rings. Some types of "moly" piston rings do not require glaze-breaking. If using these rings you should check with the manufacturer. Glaze-breaking is not recommended with some aluminum cylinder walls. Always check with the manufacturer.

Cylinder deglazing is done with a glaze-breaking tool. There are two general types of glaze breaker. The oldest type, Figure 20-5, uses three long spring-loaded abrasive stones. The newer style, shown in Figure 20-6, uses spring-loaded abrasive balls. In both, the abrasive is about 220 grit. A driver on the end of the unit is chucked in a slow-moving drill that operates at

Figure 20-6. Glaze breaker with spring-loaded abrasive balls.

Figure 20-7. A deglazed cylinder has a crosshatch pattern. (Ammco Tools, Inc.)

300 to 500 RPM. The deglazer is pushed up and down as it rotates to put the proper crosshatch pattern on the cylinder wall. The crosshatch pattern, Figure 20-7, is important because oil will cling to the small grooves created by the crosshatching. If the crosshatch pattern is too smooth, the rings will have a tough job seating. The rings need a little roughness on the cylinder walls so they will seat and wear in properly and quickly without scoring or scuffing. If the crosshatch is too rough and too deep, however, the rings will wear out too fast.

After breaking the glaze, always use soap and warm water to wash the cylinders. This combination does a good job and is the only way to get the dirt and grit out of the tiny crevices that remain after honing. Soap surrounds the dirt and grit, floating them out with water. Kerosene is not used because it drives the particles back into the crevices. Grit then acts as a grinding compound and could wear out the new rings. After cleaning the cylinders, wipe them with an oiled rag to prevent rusting.

HONING

If cylinder taper or out-of-round is beyond the manufacturer's wear specifications, the cylinder may be honed. Honing is done with a rigid honing fixture with abrasive stones, Figure 20-8, driven by an electric drill motor or drill press. This operation will remove metal from the cylinder in order to straighten it. The block usually is mounted to a plate or to the table of a drill press, Figure 20-9.

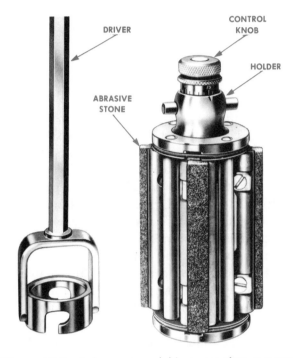

Figure 20-8. A hone uses rigid stones that remove metal from the cylinder. (Ammco Tools, Inc.)

Figure 20-9. The cylinder block is mounted to a plate or drillpress table.

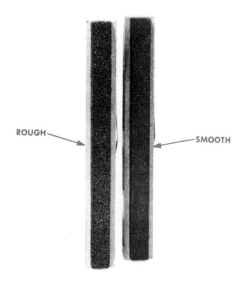

ROUGH — SMOOTH

Figure 20-10. Honing stones come in different grits or grades of smoothness.

The quality of the cylinder wall finish is affected by a number of variables. First, there are several different grits of honing stones available. Stone grits range from very rough, 70-grit, to very smooth, 600-grit, Figure 20-10. The piston ring manufacturer usually specifies what finish is best for any particular ring. A 70-grit stone provides a microfinish between 85 and 105. A 600-grit stone provides a 3 to 5 microfinish.

The speed at which the stones are rotated and pushed up and down affects the finish. A slow moving 1/2 inch drill moved up and down provides the best results.

A coolant must be used during honing. Squirt kerosene or honing oil from a squirt can onto the rotating stones. It is desirable that the stones break down during the honing process to expose a continuous supply of sharp, clean cutting edges. Honing oil will flush the loose abrasive and metal particles from the stones and cylinder wall. Equally important, it will cool the work and keep the stones clean and cutting freely. Surface temperatures are greatly reduced by the use of coolant. Coolant prevents stones from becoming loaded with particles which would cause many of the small cutting edges to "push" rather than cut. This would result in a cylinder wall finish having deep scratches, smeared grooves, torn and fragmented material, and glazed burnished areas.

Begin the honing operation in the bottom of the cylinder with a firm cutting pressure. Adjust the

Figure 20-11. Honing pressure is adjusted with a control knob on the hone.

pressure with a control on the honing fixture, Figure 20-11. Take care not to stroke the unit too far down the cylinder or the stones will not be supported properly and the finish will be poor. Take care that the stones do not come out of the top of the cylinder under power. The stones are not locked into the holder and will fly out, possi-

Figure 20-12. The hone or glaze breaker is driven with an electric drillpress.

bly injuring someone. Always wear eye protection while honing. On the drill press, a stop allows you to set the travel of the hone.

As the cylinder is machined and the stones wear, expand the stones out gradually, Figure 20-12. Wipe and measure the cylinder constantly. As soon as the bottom measurement is close to the top measurement, stroke the stones through the entire length of the bore.

Stroke the hone up and down to produce a crosshatch pattern just like that for deglazing. Take measurements often to make sure the smallest amount of stock is removed to straighten the cylinder.

The hone also may be used to machine a cylinder oversize. If the taper and out-of-round are beyond the limits specified by the manufacturer, the cylinder is honed oversize for larger pistons. Most engine manufacturers provide pistons that are larger than standard in steps of .010, .020 and .030. Choose which oversize best suits the engine. Purchase and measure the new piston. Then hone the cylinder to the new piston size plus an additional amount for clearance. When honing for oversize, stroke the entire length of the bore, to avoid honing a taper.

Cleaning is very important after the honing operation to remove abrasives and loose metal particles. Use hot, soapy water and scrub vigorously with a stiff, nonmetallic bristle brush. Scrub until the soapsuds remain white, then swab each cylinder wall with the hot, soapy solution to float out all remaining foreign matter. Next, wipe out the bore with paper toweling until clean towels show no dirt. Apply a generous coat of engine oil to all cylinder surfaces to prevent rust.

BORING

When cylinder wear or scoring is excessive, or when a perfectly straight new cylinder bore is desired, the cylinder may be bored. Boring involves machining the cylinder oversize with a cutter bit driven by a tool called a *boring bar*. After boring, the cylinder often is polished by honing to provide the desired crosshatch pattern. Boring leaves the cylinder oversize and requires that a new oversize piston be fitted.

A typical small-engine boring bar is shown in Figure 20-13. There are many different manufacturers of portable boring bars. Each type of bar is mounted and operated differently, and each manufacturer supplies step-by-step instructions for use. In this section only general procedures used with all types of boring bars will be presented.

The block must first be prepared for boring. Align the boring bar with the cylinder, using the deck surface. The surface must be clean and free of any nicks or burrs. If necessary, the surface may be cleaned up by draw filing. Position the block solidly on the boring bar table so it will not rock or vibrate during boring.

Center the boring bar over the cylinder to be bored. Centering the bar is very important to insure that the newly machined cylinder is in the same position as the old one. The boring bar has a handwheel which allows the operator to feed the bar down into the cylinder with the power off. Most boring bars have three centering fingers or plungers on the end of the bar. Turning a control knob pushes these three fingers out of the bar. The fingers contact the cylinder wall at three equidistant spots. As the fingers are brought out tightly against the cylinder, position the block in the exact center of the boring bar.

For normal boring operation, the centering is done in the unworn lower part of the cylinder. This provides the most accuracy in centering. The area used for centering must be perfectly clean. With the cylinder centered, clamp the block securely in place on the table. After centering and anchoring, raise the boring bar out of the cylinder.

The next step is to prepare the cutter bit or tool bit. The tool bit does the actual machining. It is a

Figure 20-13. A small engine block mounted to a boring bar. (Kwik-Way)

small bar of tool steel with a cutting edge shaped on the end. Many tool bits have a tungsten carbide tip on the cutting end. Sharpen the cutting tip each time it is used. The tool bit is sharpened on an iron disc that has its surface charged with diamond dust. The disc usually is mounted on

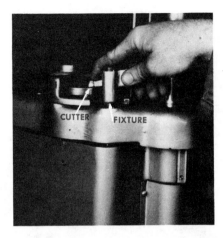

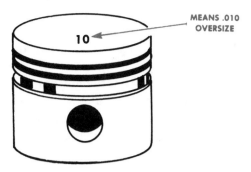

Figure 20-15. A number on top of the piston means it is oversize. (Clinton Engines Corp.)

Figure 20-14. The cutter bit is sharpened in a fixture on the boring bar. (Kwik-Way)

and driven by the boring bar motor. Carefully maintain the angles on the tip of the tool bit for proper cutting. Install the cutter in the special fixture which is installed on a pilot shaft over the disc, Figure 20-14. The fixture controls the position of the tool bit as it is sharpened.

After the tool bit is sharpened, it is ready to be installed in the boring bar. The next operation involves setting the tool bit to machine the cylinder to the desired size. The size to which the cylinder will be bored depends upon what sizes of oversize pistons available. Choose an oversize piston that allows the cylinder to be completely remachined. For example, if cylinder wear is in excess of .010, a .020 oversize piston should be used. On the other hand, boring the cylinder .020 oversize when .010 would be enough takes away the possibility of another rebore when the engine is worn again. Piston oversizes are marked on the top of the piston. A 10 or .010 on the top of the piston means it is .010 over standard, Figure 20-15.

The usual practice is to purchase and have on hand the piston to be used. Then measure the diameter of the new piston. Set the boring bar to machine the cylinders at exactly this measurement. Achieve the proper clearance between the cylinder and piston by polishing the rebored cylinder with a hone.

The diameter to which the cylinder is machined is determined by the position of the tool bit in the

Figure 20-16. The size to which the cutter bit machines is set with a micrometer on the boring bar. (Kwik-Way)

boring bar. The position of the tool bit is set accurately with a special micrometer, Figure 20-16, that is part of the boring bar tool kit. In many types of boring bars, the tool bit fits in a holder that in turn fits into the boring bar. To install the tool bit into the holder use a micrometer adjusted to the desired dimension of the cylinder. Adjust the tool bit and holder to this dimension. The holder is then mounted into the boring bar. On other types of boring bars, the micrometer is designed to adjust the tool bit when it is in the boring bar.

With the tool bit properly set, the machining operation can begin. Turn on the motor to rotate the bar and tool bit. Engage the feed lever to cause the bar to move slowly down the cylinder. The rotating and advancing tool bit machines the cylinder. Motor speed and feed speed controls may be used to vary the quality of the finish.

When the tool bit has advanced through the bottom of the cylinder, switch off the bar. Remove the tool bit from the bottom to prevent scratching the new cylinder as the bar is raised out of it. A special long tool bit is then used to put a slight chamfer in the top of the cylinder. This helps get the piston and piston rings into the cylinder.

Boring bars and boring bar operators differ in their ability to get a good cylinder wall finish. It is possible to get a very highly finished cylinder by boring. Even so, most mechanics prefer to polish the rebored cylinder with a hone. Honing is the only way to develop a crosshatch pattern. The pattern developed by a tool bit going around and down the cylinder looks like a thread. This pattern does not hold lubricant as well as a crosshatch does.

The cylinders are honed using polishing grit stones. Just enough material is removed to establish the specified amount of clearance between the piston and cylinder wall. This procedure is called *fitting* the pistons.

PISTON INSPECTION AND MEASURING

When cylinders are rebored, use new pistons. The only service required on new pistons is careful measurement and fitting of the new cylinder. If the cylinders are not rebored, the original pistons must be serviced.

The first step in piston service is to check each of the pistons visually. Carefully check for fractures at the ring lands, at the skirts and pin bosses.

SCORE MARKS

Figure 20-17. Pistons with signs of scoring or scuffing should be replaced.

Figure 20-18. Pistons are measured across the skirt at right angles to the pin hole.

Look for scuffed, scored or rough surfaces, Figure 20-17. Get new pistons for those that show signs of too much wear or have wavy right lands.

If the pistons pass a visual inspection, measure them with a micrometer. Measure the outside diameter of the piston at the piston skirt, at 90 degrees to the piston pin, Figure 20-18. If the piston dimension meets the manufacturer's specifications, continue to inspect it.

Check the top piston ring groove for wear. Place a new compression ring in the groove, as shown in Figure 20-19. It is not necessary to install the ring at this time. Use a feeler gage as shown and run the ring all the way around the groove. If the clearance is over manufacturer's specifications, use a new piston.

The top ring is the most important ring on the piston. It acts as a compression ring to control combustion pressures and also as a final oil control ring. The efficiency of the ring depends on the ability of the ring face and sides to form effective seals with the cylinder wall and with the sides of the ring groove. For maximum performance, the sides of the ring grooves must be flat, parallel and smooth.

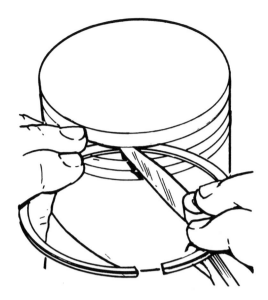

Figure 20-19. Piston ring groove wear is checked with a new ring and a feeler gage. (Onan Corp.)

The top ring and top groove of most aluminum pistons that have been in service for a long time become worn due to abrasives and high temperature in the top ring land area. This increases the top ring side clearance and causes increased blow-by and oil consumption.

If a new top compression ring is installed in a worn groove, a proper seal cannot be formed. The worn groove forces the upper outside edge of the ring face to contact the cylinder wall, causing the oil to be wiped up into the combustion chamber instead of down into the crankcase. In addition, the continued twisting of the new top ring in the worn groove will result in ring breakage.

MEASURING RING END-GAP

Before installing new rings on a piston, push down each compression ring from the ring set to the lower, unworn portion of a cylinder. The ring must be square in the cylinder. Pushing the ring down in the bore with the head of the piston puts the ring into the bore at the proper depth and squares it with the cylinder walls. Use a feeler gage to measure the ring end-gap, Figure 20-20. Checking this measurement with that listed in the specifications provides a double check as to whether or not the ring set is the correct size.

PISTON RING END GAP

POSITION RING HALFWAY IN CYLINDER
CAUTION: DO NOT SCRATCH CYLINDER WALL

Figure 20-20. The ring end-gap must be checked for each piston ring. (Onan Corp.)

It is important that all rings have at least the minimum gap. This is necessary to provide for the difference in expansion which may occur between the piston ring and the cylinder. Otherwise, the ring ends may butt and cause scuffing, scoring, ring breakage or engine seizure. End clearance less than the required minimum is caused by the cylinder being undersize or by using the wrong ring set. Too much end-gap can cause leakage.

NEW TERMS

boring: Machining away metal from a worn cylinder with a boring bar for the installation of new oversize pistons.

counterbored ring: A ring with a section removed to allow twisting and improvement of sealing.

cylinder gage: A dial gage made to measure cylinder wear.

cylinder out-of-round: A wear condition in a cylinder in which the cylinder is worn more in one direction than another.

cylinder taper: A wear condition in which the top of the cylinder is larger than the bottom.

expander: A flexible spring placed behind a piston ring to increase spring tension.

glaze-breaking: Roughing up the surface of a cylinder with an abrasive stone to help the rings seat.

honing: Removing metal from a cylinder with abrasive stones to straighten it.

piston ring end-gap: The space between the ends of a piston ring when they are in the cylinder.

piston ring side clearance: The space between a piston ring and ring land.

SELF CHECK

1. What are two kinds of wear common to cylinders?
2. Describe how to measure a cylinder to determine taper.
3. What is glaze-breaking?
4. When is a cylinder honed?
5. Describe how to hone a cylinder.
6. Why is it necessary to use oversize pistons after boring?
7. Describe how to measure the diameter of a piston.
8. How are piston ring grooves measured for wear?
9. How is piston ring end-gap measured?

DISCUSSION TOPICS AND ACTIVITIES

1. Use a practice cylinder block and measure cylinder taper and out-of-round. Decide how the cylinder could best be reconditioned.
2. Use practice parts to measure piston diameter, ring side clearance and ring end-gap.

unit 21
engine reassembly

When all the engine machine work is completed and all the new parts and gaskets are on hand, the engine reassembly can begin. One of the most critical factors in engine overhaul is preventing dirt and other foreign material from entering the engine during assembly. You must work carefully to insure a clean engine. The reassembly must be done as far away from valve grinding and machine work as possible. As parts are installed, wipe them carefully and inspect for dirt. Clean tools carefully before they are used. Any time the engine is not being worked on, it should be covered. A large plastic trash bag may be pulled over the block assembly and sealed closed with masking tape. In this unit we will see how the engine is reassembled.

LET'S FIND OUT: **When you finish reading and studying this unit, you should be able to:**
1. **Describe how to install piston rings on a piston.**
2. **Explain how to install a piston in a cylinder.**
3. **Describe how to install the crankshaft and camshaft.**
4. **Explain how to measure connecting rod oil clearance.**
5. **List the steps to follow in installing cylinder heads and accessories.**

GASKETS AND SEALS

When an engine is overhauled, all the gaskets and seals are replaced. Gaskets for an engine may be purchased individually or in a set. A complete engine overhaul requires an overhaul gasket set, which includes all of the engine gaskets and seals. An overhaul gasket set is shown in Figure 21-1.

Gaskets are used between two parts that are bolted together, to form a pressure-tight seal. Gaskets are necessary in most places because

Figure 21-1. An engine overhaul requires a complete gasket set.

233

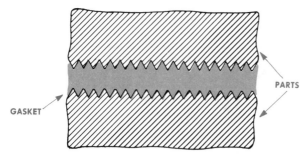

Figure 21-2. A gasket compresses into irregularities between parts to form a seal.

there are microscopic irregularities on the component surfaces. If the parts were fastened together without a gasket, the high and low spots on the parts would allow leakage. A gasket is a soft material which, when squeezed between two components, fills up these small irregularities and forms a pressure-tight seal, Figure 21-2.

Sealing around a rotating or sliding shaft cannot be done with a gasket. This is the job of special parts called *seals*. There are many different types and designs of seals but the lip type is the most common in small engines. The lip seal, Figure 21-3, is used to seal a lubricant inside a bearing area, and to keep dirt or other abrasive materials out of that area. A lip seal often is used around the crankshaft in the block and side cover. The actual sealing job is done by only a very small part of the sealing element. The knife-sharp edge of the lip

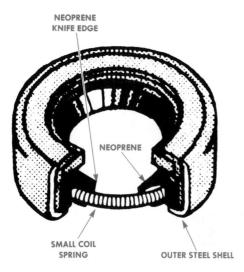

Figure 21-3. The main parts of a seal. (Clinton Engines Corp.)

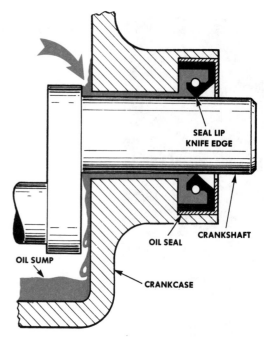

Figure 21-4. The seal lip knife edge wipes the crankshaft to prevent oil leakage.

hugs the shaft and acts as a squeegee, wiping the lubricant from the shaft and preventing its escape, as shown in Figure 21-4. Despite any differences in the appearance and construction of lip seals, they all work in the same way.

A new seal must be installed in the block and side cover. Remove old seals by prying them out with a seal remover pry bar. Drive the new seal into the block or side cover using a block of wood, as shown in Figure 21-5. Make sure the knife edge of the seal is toward the inside of the engine.

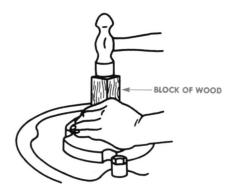

Figure 21-5. Installing a crankshaft seal with a block of wood prevents damage to the seal.

GASKET INSTALLATION PRECAUTIONS

Gaskets should be handled with care to avoid distorting or bending them. Try to store gaskets flat, not on edge, and keep them in cartons until ready for use. A distorted or bent gasket may look satisfactory but it may fail prematurely. Never rinse gaskets. A new gasket conforms to microscopic irregularities in the parts between which it is clamped. The gasket material between the parts is permanently compressed. It is impossible to replace the gasket in its exact original position, and the compressible material will not readjust itself to surface irregularities.

Most important is a thorough cleaning of the cylinder head, block, bolts and bolt holes. A dirty bolt can throw off torque readings 20 inch-pounds or more. Bolt threads in the block frequently are overlooked. Dirt particles form pockets in the gasket, allowing leakage and eventual gasket failure. Inspect bolt holes carefully, especially if the surfaces have been refaced. If the threads run up to the surface, chamfer the surface and retap the hole so that threads will not be drawn above the surface when bolts are tightened. This slight irregularity is enough to cause early gasket failure.

Choose the correct gasket for the job you are doing and then check to see that it fits. Assuming the fit to be correct either way, a cylinder head gasket should function equally well whether installed face up or inverted. However, some gaskets may be labeled "This side up" or "Front" to assure correct operation of the engine. If so, follow the instructions on the gasket.

INSTALLING THE PISTON ON THE CONNECTING ROD

Engine reassembly begins with the installation of the piston on the connecting rod. In many four-stroke engines, the direction of the piston is not important. Two-stroke-cycle pistons that have contoured tops, Figure 21-6, must be installed correctly for the engine to run. Observe the marks made on the piston and rod during disassembly.

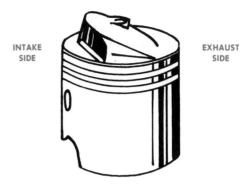

Figure 21-6. Two-stroke pistons must be installed in the correct direction. (Clinton Engine Corp.)

Take care when removing or installing the wrist pin in the rod or piston. It is easy to distort or damage piston or rod when installing the wrist pin. Never lay the piston on a solid object when removing or installing the wrist pin. To prevent damage, support the piston in the palm of your hand when servicing. In most small engines, there is no special way to install the wrist pin. Some two-cycle-engines, however, have a hollow wrist pin closed on one end. In this case, make sure the closed end is toward the exhaust side. Push the pin through the piston and rod, Figure 21-7. If it sticks, heat the parts under hot water. It is good practice to use new retaining rings. The old ones often are stretched out of shape during disassembly and may not stay tightly in their grooves.

Figure 21-7. The pin is pushed into place by hand. (Briggs & Stratton Corp.)

INSTALLING PISTON RINGS

If the ring end-gap is satisfactory, the rings may be installed on the piston. Most compression rings must be installed with their top side toward the top of the piston in order to work correctly. Diagrams such as that in Figure 21-8 are provided with new piston ring sets to show which ring goes in which groove and in what direction each must be installed.

When installing rings, use a good quality ring-expander tool to prevent over-spreading the rings, Figure 21-9.

In most installations, ring ends may be spaced equally at 120° intervals around the piston. Some

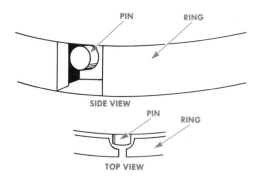

Figure 21-10. Two-stroke pistons often have pins to prevent rings from turning. (Clinton Engines Corp.)

manufacturers specify a particular spacing for the ring ends. Many two-stroke engines use a pin in the ring groove to prevent the rings from rotating, Figure 21-10. The ends of the ring must match up with the pin.

INSTALLING THE CAMSHAFT AND CRANKSHAFT

Oil the crankshaft and slide it into the main bearing in the block. Install the valve lifters and position the camshaft in the bearing. The crankshaft and camshaft gears must be meshed correctly to time the camshaft to the crankshaft. Move the camshaft around until the timing marks are properly aligned, as shown in Figure 21-11. If the timing is off, the engine will not run.

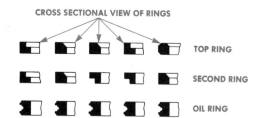

Figure 21-8. Instructions with the rings show how they are installed. (Briggs & Stratton Corp.)

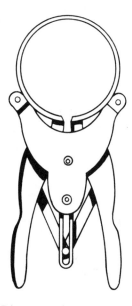

Figure 21-9. Rings are installed with an expander tool. (Ammco Tools, Inc.)

Figure 21-11. The timing marks on the camshaft and crankshaft must be aligned correctly.

If the crankshaft and camshaft are not installed correctly in relation to each other, the valves will not open or close at the correct time in relation to the piston position. If, for example, the piston moves down on an intake stroke, but the intake valve is not open, air and fuel cannot enter the cylinder. Similarly, the other strokes, compression, power and exhaust, will not operate correctly and the engine will not run.

INSTALLING THE PISTON ASSEMBLY

The piston and connecting rod assemblies may be installed in the cylinder block next. A new bearing insert, when used, is used in each of the connecting rods. The piston rings and cylinder walls are coated with oil. This lubricates piston rings and cylinders during the engine cranking period and until the oil throw-off from the connecting rod journals is adequate. The piston can be submerged in a can of clean engine oil just before it is installed.

Use a piston ring compressing tool, Figure 21-12, to install the piston and ring into the cylinder bore. The tool is a thin metal band that is tightened around the piston head to squeeze the rings tightly into their grooves. The tool is expanded and contracted with an allen wrench. There are

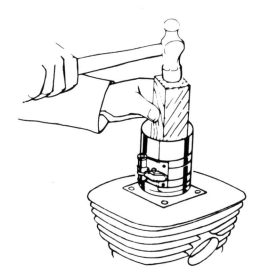

Figure 21-13. A hammer handle or block of wood is used to tap the piston through the compressor and into the cylinder. (Ammco Tools, Inc.)

some small steps on the bottom of the tool that prevent it from entering the cylinder.

After the piston is oiled, expand the compressing tool and place it around the piston rings. Position the steps on the tool downward. Tighten the tool with the wrench to compress the piston rings. When the rings are fully compressed, the tool will not compress any further.

Insert the piston and connecting rod in the correct direction into the cylinder bore until the steps on the compressing tool contact the cylinder block deck. When the rings are properly compressed into their grooves, only a light tapping on the piston head with an average-size hammer handle or block of wood is required to push the piston assembly into its bore, as shown in Figure 21-13. Make sure the rod is installed properly. Oil holes in the rod must face in the correct direction or the connecting rod bearing may not get enough oil. Observe marks made during disassembly.

INSTALLING THE CONNECTING ROD CAP

With the crankshaft installed, the rod cap is ready for installation. The rod cap must be installed in the correct direction. Observe the

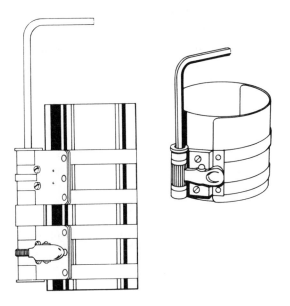

Figure 21-12. A ring compressing tool is used to install the piston assembly. (Ammco Tools, Inc.)

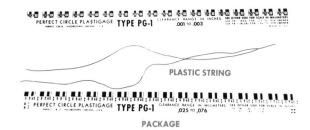

PLASTIC STRING

PACKAGE

Figure 21-14. Connecting rod bearing clearance is determined with plastic string.

marks made on the rod and rod cap during disassembly.

Measure the oil clearance between the crankshaft and the connecting rod bearing. Correct crankshaft oil clearances are necessary to allow for proper lubrication and cooling of the bearing during operation. If the oil clearance is too small, many problems such as wiped bearings, worn crankshaft, excessive cylinder wear, stuck piston rings and worn pistons may result.

If the oil clearance is too great, the crankshaft may overheat and weld itself to the insert bearings. A specially molded, fine plastic string is used to determine the oil clearance between the bearing and the shaft, Figure 21-14. The string is small in diameter, relatively long and of soft plastic. A length of it is cut off approximately ⅛ inch shorter than the bearing and placed on the bearing inside surface or on the shaft after each has been wiped free of oil.

Assemble the rod cap to the connecting rod. Tighten the rod cap bolts to the correct torque specifications. Be careful not to turn the crankshaft because this will ruin the plastic string.

Remove the rod cap. The plastic string will now be flattened. Compare the strips in the plastic string package as shown in Figure 21-15, until you find a strip that is the same width as the flattened string. Each stripe on the package has a measurement printed on it. The measurement printed on the matching stripe is the oil clearance.

One side of the plastic string package has clearance stripes for customary measurements. The other side has stripes for measuring clearance in metric measurements. The string may be purchased to measure different clearance ranges. Normally only the smallest clearance range is necessary for reassembly work. The string designed to measure wide clearances may be used for troubleshooting worn engines. If the clearance is not to specifications, the engine cannot be reassembled further until the cause is determined. The new bearings may be the wrong size or an error may have been made in crankshaft grinding or connecting rod reconditioning.

If the clearance is within specified limits, the connecting rod bearings should be oiled. The cap is replaced and torqued to specifications. Each

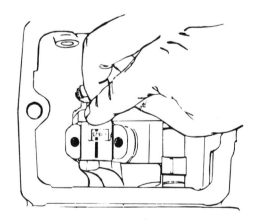

Figure 21-15. The amount the string is flattened is measured to find the clearance. (Clinton Engines Corp.)

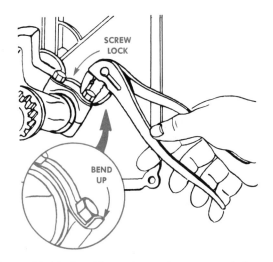

Figure 21-16. Locking tabs are bent around the connecting rod bolts or nuts. (Briggs & Stratton Corp.)

Figure 21-17. End-play is corrected by selecting a side cover gasket of the correct thickness.

cap should be checked to be sure it is on the correct rod and has not been reversed. Many rods have lock plates or tabs that must be bent up around the connecting rod nuts or bolts, as shown in Figure 21-16. These tabs are very important in preventing the nuts or bolts from vibrating loose. If the connecting rod has an oil dipper, it must be installed at this time. Make sure the dipper is installed in the correct direction or the engine will not be oiled properly.

INSTALLING THE SIDE COVER

Before installing the side cover, make a final check inside the engine. Check that the timing marks and the connecting rod cap are installed in the correct position. Make sure any oil dipper, oil slinger or governor parts are installed in the correct position.

The thickness of the side cover gasket sets the crankshaft end-play. Check the end-play specifications and compare them to the end-play measurement made during disassembly. The end-play may be increased or decreased by selecting thicker or thinner gaskets. Use an outside micrometer to measure the thickness of the old side cover gasket, as shown in Figure 21-17.

If the end-play was correct on disassembly, choose a gasket of the same thickness as the origi-

nal from the gasket set. A thicker gasket will increase end-play. A thinner gasket will reduce end-play. Place the gasket on the block. Wrap the end of the crankshaft with tape to prevent damage to the seal in the side cover as it goes over the crankshaft. Slide the cover over the crankshaft. Tighten the side cover bolts with a torque wrench. If all the parts are installed correctly, the engine should turn freely by hand. Always recheck the end-play.

VALVE ADJUSTMENT AND INSTALLATION

The next step in reassembly is to adjust and install the valves. These procedures were described in the chapter on valve service.

INSTALLING THE CYLINDER HEAD

Before installation, check the cylinder head for warpage. A warped cylinder head will not seal in compression pressures. Place the cylinder head on a flat surface such as a drillpress table. Look for light between the table and the cylinder head. Any space between the head and the table indicates warpage. In that case, the cylinder head must be resurfaced straight or replaced. To resurface, place the head on emery cloth on a flat surface and push it back and forth across the cloth.

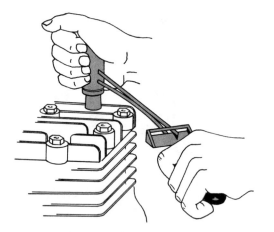

Figure 21-18. Cylinder head bolts are tightened with a torque wrench.

Place the head gasket on the block in the correct position. Set the cylinder head on top of the gasket. Start each cylinder head bolt by hand. Most cylinder heads use both long and short bolts. Check the service manual or disassembly notes for the positions of the long and short bolts. Tighten the cylinder head bolts in the correct order and to the correct torque with a torque wrench, Figure 21-18.

FINAL REASSEMBLY AND START-UP

Install all the accessories on the outside of the engine. If any service is required on the fuel or ignition system, it should be done at this time. Install the engine on the equipment or vehicle. Fill the crankcase with the correct amount of recommended oil. Fill the fuel tank with fresh fuel. Start the engine and allow it to run at fast idle for a few minutes. The engine will smoke for a short time until oil on the cylinders and rings burns off. Make the final carburetor adjustments described under fuel system service.

Initial oil consumption in new engines or after installation of new rings is often higher than normal for a short period of time until the rings become seated. That period of engine operation which is required before blow-by and oil consumption drop off to an acceptable level is referred to as the *break-in period*. Different piston ring manufacturers specify different break-in

procedures. Follow the one required for the rings that were installed.

Ring seating consists of the mating of the ring face with the cylinder wall throughout the complete stroke of the ring. This is accomplished as the very slight irregularities of the ring face and the cylinder wall are worn off. The break-in period has been reduced as rings and cylinders are produced with greater accuracy, but, even with the finest equipment, ring and engine manufacturers have not been able to duplicate the mating surfaces made by the engine. For this reason, all-new, rebuilt or engines with new rings require a certain amount of running to obtain maximum oil control.

NEW TERMS

break-in: The wearing-in to improve the fit between the piston rings and cylinder walls.

connecting rod bearing clearance: The space for oil between the crankshaft and the connecting rod bearing.

cylinder head warpage: A condition in which the cylinder head is not flat.

gasket: A soft material used between two engine parts to form a pressure-tight seal.

plastic string: A string used to determine bearing clearance.

seal: A device with a flexible lip used to keep lubricant inside a bearing area.

SELF CHECK

1. What is the purpose of a gasket?
2. How does a seal work to prevent oil loss?
3. Why should gaskets not be reused?
4. Describe how to assemble a piston to a connecting rod.
5. Explain how to install piston rings on a piston.
6. Why must the crankshaft and camshaft be properly timed?

7. What is a piston ring compressor?
8. Describe how to measure connecting rod bearing oil clearance.
9. How is crankshaft end-play adjusted?
10. How is the cylinder head checked for warpage?

DISCUSSION TOPICS AND ACTIVITIES

1. Make an outline of steps to follow in assembling an engine.
2. Look up the torque specifications for an engine you own.

Outboard boat engines may be
considered small engines. (Evinrude Motors)

part 7
specialized engine service

As a small engine technician you may work just on single-cylinder lawn or garden equipment engines. You may, however, wish to branch out into more specialized types of engines. Two areas of specialized engine repair open to small-engine technicians are motorcycle and outboard. The motorcycle service area includes work on mopeds and snowmobiles. These vehicles are becoming more popular because they use very little gasoline. Outboard engines have always been popular in areas around lakes, rivers or the ocean. In this part we will study these engines in more detail.

unit 22

motorcycle, moped and snowmobile engines

One specialized area of engine service for the small-engine technician includes motorcycles, mopeds and snowmobiles. These vehicles use engines that work on the same principles already presented. They may be four- or two-stroke-cycle engines. Many of these engines have more than one cylinder, but the parts are much the same as those in the single-cylinder engines we have studied. These engines have ignition, fuel, lubrication and cooling systems similar to those on single-cylinder engines. We use the same trouble-shooting, maintenance and tuneup procedures on these engines as we do on the single-cylinder engines. In this unit, we will study the parts and repair procedures that are different from those in single-cylinder engines.

LET'S FIND OUT: **When you finish reading and studying this unit, you should be able to:**
   1. **Describe how a two-stroke motorcycle engine works.**
   2. **Explain the operation of a rotary valve.**
   3. **Recognize the main parts of a motorcycle engine.**
   4. **Describe how to service a sliding-valve carburetor.**
   5. **Explain how to service a multi-piece crankshaft.**

ENGINE OPERATION

Many motorcycles, Figure 22-1, and mopeds, Figure 22-2, have engines that work on the two-stroke-cycle principle. These engines are generally the same as other two-stroke engines we have studied. The main difference is that a transmission is built into the crankcase.

A motorcycle or moped engine has three main parts, as shown in Figure 22-3. The cylinder section, or upper end, houses the piston and cylinder head which are the same as most other two-stroke pistons and cylinders. The crankcase section, as in

Figure 22-1. Many motorcycles use a two-stroke engine. (Suzuki Motor Co.)

Figure 22-2. Mopeds like this one have a two-stroke engine. (Batavus USA)

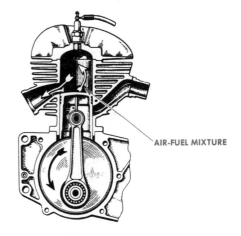

Figure 22-4. Transfer or intake-to-cylinder phase. (Yamaha Motor Corp., USA)

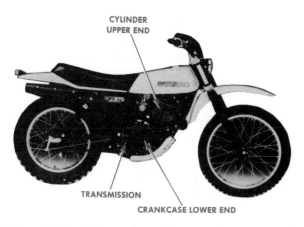

Figure 22-3. Main sections of a moped or motorcycle two-stroke engine.

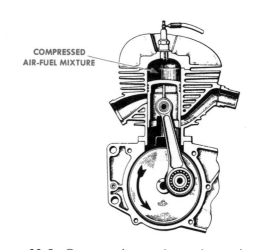

Figure 22-5. Compression and crankcase intake phase. (Yamaha Motor Corp., USA)

other two-stroke engines, supports the crankshaft. This is often called the *lower end*. Since the engine is used to drive the rear wheel of the moped or motorcycle, a clutch and transmission are necessary. The parts of the transmission and clutch fit into the crankcase of the engine.

The operation of the two-stroke moped or motorcycle is similar to the operation of other types of two-stroke engines. During the transfer phase, Figure 22-4, the piston moves down the cylinder. Air and fuel in the crankcase are pressurized because the piston makes the crankcase area smaller. The air-fuel mixture cannot escape out of the crankcase anywhere except through the

transfer port, so it goes up the transfer port and into the cylinder through the open intake port.

As the piston moves back up the cylinder, the compression and crankcase intake phase begins, Figure 22-5. The piston covers up the intake and exhaust ports. The air-fuel mixture is trapped above the piston. As the piston moves up, it compresses the mixture. At the same time, the piston movement upward creates a low pressure area in the crankcase. This low pressure pulls a new charge of air and fuel into the crankcase.

Just as the piston reaches the top of its travel, a spark is created at the spark plug. The air-fuel

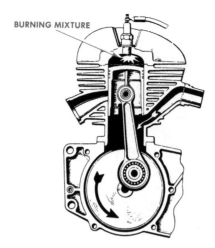

Figure 22-6. Power phase. (Yamaha Motor Corp., USA)

mixture is ignited. The pressure from the burning mixture pushes the piston down the cylinder. This is called the *power phase*, Figure 22-6. While the power phase is occurring, the transfer phase is also starting again. Another charge of air-fuel mixture begins up the transfer port. The charge cannot enter the cylinder yet because the piston is still covering the port.

When the piston has moved far enough down the cylinder, it will begin to uncover the exhaust ports. The exhaust ports are located at a point above the intake ports. As soon as the port is uncovered, the high pressure in the cylinder pushes out the exhaust gases, Figure 22-7. Further

movement of the piston downward uncovers the intake port and a new charge enters the cylinder. All phases of operation have occurred in two piston strokes or one crankshaft revolution.

ROTARY VALVE

Many moped and motorcycle engines control the flow of air-fuel mixture into the cylinder with a reed valve. The reed valve operates just like those described earlier.

Some engines, instead, use a device called a *rotary valve* to control the flow. A rotary valve is shown in Figure 22-8. The valve is mounted to the crankshaft. The shape of the valve is such that during part of the crankshaft revolution it is open and allows air and fuel to enter the vacuum of the crankcase, Figure 22-9. As the crankshaft rotates

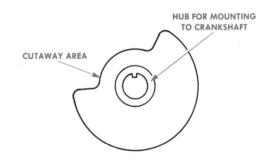

Figure 22-8. Rotary valve.

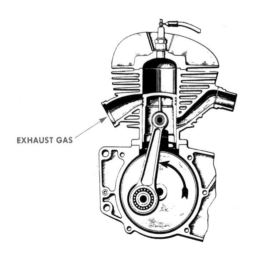

Figure 22-7. Exhaust phase. (Yamaha Motor Corp., USA)

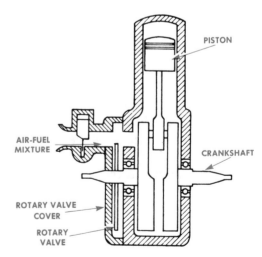

Figure 22-9. The rotary valve is turned so that it is open.

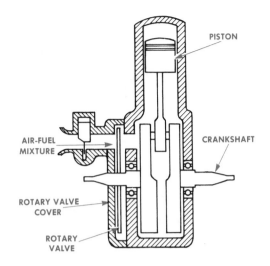

Figure 22-10. The rotary valve has turned to close off the port.

further, the valve closes off the port, Figure 22-10. The port is closed when the crankcase is under pressure.

ENGINE PARTS

Most of the other engine parts for a moped or motorcycle two-stroke engine are similar to those in other two-stroke engines. There are, however, some differences in cylinder head and crankshaft design. The cylinder heads often are designed with large cooling fins to help remove the heat. These cylinder heads, Figure 22-11, are usually called sunburst heads.

Some two-stroke engines use cylinder heads with two spark plugs, as shown in Figure 22-12. When one plug fouls out, the ignition may be connected to the other plug. Other cylinder heads are equipped with a valve, Figure 22-13. The valve

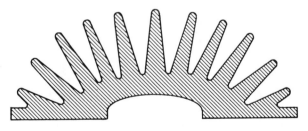

Figure 22-11. Many motorcycles use a sunburst cylinder head.

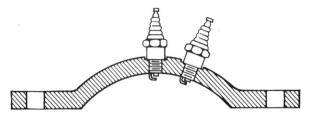

Figure 22-12. Cylinder head with two spark plugs.

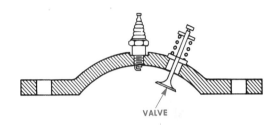

Figure 22-13. Cylinder head with a compression release valve.

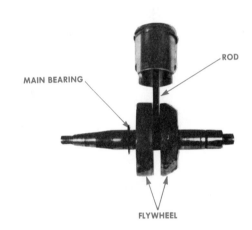

Figure 22-14. The crankshaft and flywheel are combined in many motorcycle engines.

is called a *compression release.* When the rider pushes it down, the engine may be turned over without any compression resistance. This allows the engine to be turned over rapidly to dry out a cylinder that may be wet with fuel.

The lower end of a motorcycle or moped engine may use a combination crankshaft and flywheel assembly, Figure 22-14. The counterweights on the crankshaft are two large wheels which take the place of the flywheel. The connecting rod typically does not have a cap. The rod is constructed

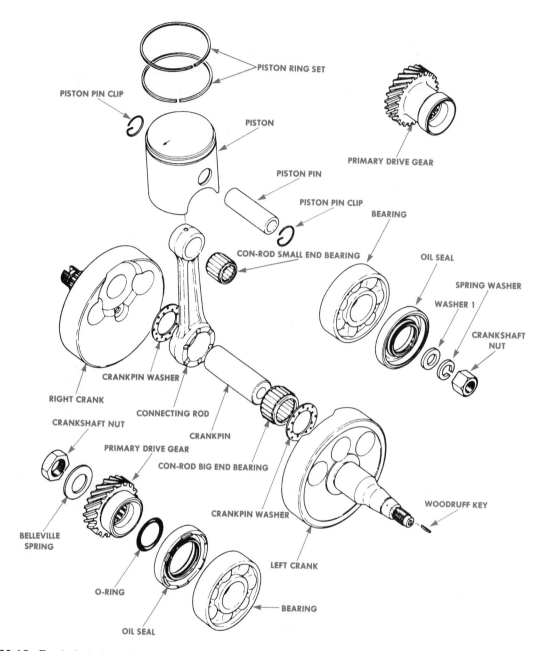

Figure 22-15. Exploded view of crankshaft assembly. (Yamaha Motor Corp., USA)

in a full circle. The rod is installed by pressing the crankpin out of the crankshaft. Slide the rod and bearing over the crankpin and press the crankshaft back together. An exploded view of the crankshaft ports is shown in Figure 22-15.

MOTORCYCLE CARBURETOR SERVICE

Many motorcycles use a sliding-valve carburetor. It is serviced during a fuel system tuneup. The following procedures may be used along with the manufacturer's service manual to repair a sliding-valve carburetor.

Clean all around the carburetor. With the fuel valve lever placed in "OFF" position, disconnect the fuel line from the carburetor. Remove the inlet pipe one-way bolt, Figure 22-16. Do not allow dust and dirt to come in contact with the bolt. Take care not to allow the oil to come out of the outlet hose.

Remove the inlet pipe tightening bolt. Then remove the connecting tube band and remove the carburetor and inlet pipe together from the cylinder and air cleaner connecting tube, Figure 22-17. Then plug the cylinder and air cleaner to keep out dust and dirt.

Loosen the carburetor top and remove it together with the throttle valve, Figure 22-18. Put the throttle valve in a plastic bag to prevent dust and dirt from coming in contact with it. Remove the insulator band and remove the carburetor from the inlet pipe.

Remove the needle clip plate from the throttle valve. Take the valve plate out of the throttle valve and pull the throttle cable out of the groove

Figure 22-16. Preparing to remove carburetor.

Figure 22-17. Removing the carburetor.

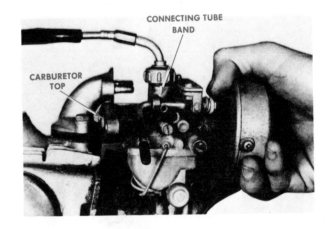

Figure 22-18. Removing carburetor top.

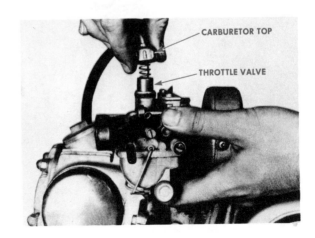

Figure 22-19. Removing sliding valve assembly from carburetor.

in the throttle valve. At this time, take care not to let the jet needle come out accidentally. Remove the rubber cap and disconnect the throttle cable from the carburetor top, Figure 22-19.

Use an exploded view of the carburetor, Figure 22-20, as a guide to disassemble it.

Remove the four phillips screws holding the float bowl to the body. With the carburetor upright, remove the float bowl.

Carefully set the body aside and inspect each independent float within the float bowl cavity. Note where each one is installed. The float arm pin must be on the lower side of the float and toward the center. Inspect the float. If fuel has

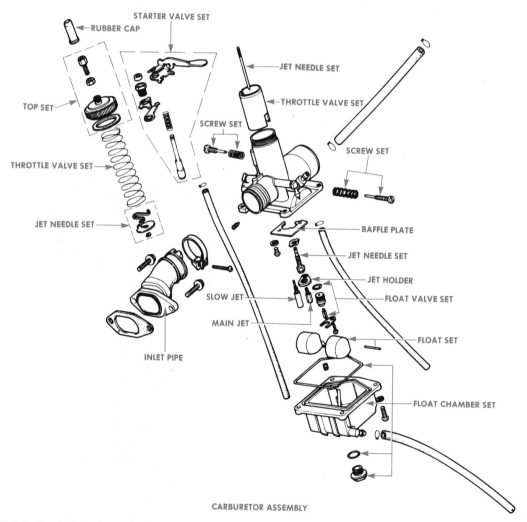

CARBURETOR ASSEMBLY

Figure 22-20. Exploded view of sliding valve carburetor. (American Honda Motor Co., Inc.)

entered it, the float must be replaced. On the carburetor body, remove the pin securing the float arm. Remove the arm.

Remove the inlet needle directly beneath the float arm tang. Inspect the needle and seat for signs of excessive wear or attached foreign particles. Replace as required. Always replace the inlet needle and inlet valve seat as an assembly. Then remove the main jet and slow or idle jet. Soak the carburetor parts in a cold tank. Flush the parts with water and blow out all passages and parts with compressed air. Blow the jets to check for clogging. Check the float valve for proper operation. Check the throttle valve for breakage or wear. Check the jet needle for breakage or wear.

Begin the reassembly of the carburetor by installing the parts in the carburetor body. After the float has been installed, adjust the float level. Hold the carburetor with its main bore in a vertical position, so the float arm tang will just close

the float valve, without compressing the spring-loaded plunger in the end of the valve. Measure float height with a float-level gage, Figure 22-21. Float height (distance between the carburetor body and the opposite edge of the float) should be set to manufacturer's specifications when the float valve just closes. If adjustment is needed, carefully bend the float arm tang toward or away from the float valve until the specified float height is obtained.

Complete the reassembly of the carburetor. The unit may then be reinstalled on the engine. Check the hand throttle for proper operation. Turn on the fuel valve and check for fuel leaks. Follow the manufacturer's recommended procedure for final adjustments.

ENGINE SERVICE

Many moped and motorcycle engines use a multi-piece crankshaft. The connecting rod journal is pressed into the counterweights between the main bearing journals. A damaged connecting rod journal is replaced by pressing the crankshaft apart and then installing a new connecting rod journal, as shown in Figure 22-22. The crankshaft parts are assembled and pressed together in a special holding fixture to make sure that they are in correct alignment.

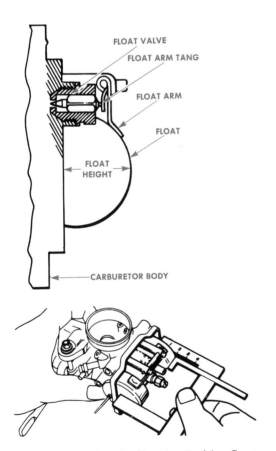

Figure 22-21. Setting the float level with a float gage. (American Honda Motor Co., Inc.)

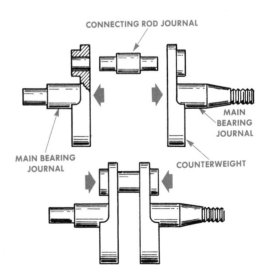

Figure 22-22. A new connecting rod journal can be installed on a multi-piece crankshaft.

NEW TERMS

compression release: A valve in the cylinder head that allows compression pressure in the cylinder to be released.

float-level gage: A tool used to measure the float level in a sliding-valve carburetor.

lower end: The crankshaft area of a moped or motorcycle engine.

rotary valve: A crankshaft-mounted valve that controls air-fuel mixture flow into a motorcycle engine.

upper end: The cylinder area of a moped or motorcycle engine.

SELF CHECK

1. List the three main parts of a motorcycle or moped engine.
2. Explain what happens in a motorcycle engine when the piston moves down.
3. Explain what happens in a motorcycle engine as the piston moves up.
4. What is the purpose of the rotary valve?
5. How does the rotary valve work?
6. List the steps in removing a sliding-valve carburetor.
7. What parts are replaced when you service a sliding-valve carburetor?
8. How is the float measured and adjusted in a sliding-valve carburetor?
9. List the steps in disassembling a sliding-valve carburetor.
10. How is a multi-piece crankshaft serviced?

DISCUSSION TOPICS AND ACTIVITIES

1. Using a motorcycle, moped or snowmobile, make a list of all the parts you can identify.
2. Using an owner's manual for a motorcycle, moped or snowmobile, make a list of the maintenance operations necessary to keep the machine operating properly.

unit 23
outboard engines

Many small-engine technicians work on outboard marine engines. These engines work on the same principles and have many of the same parts as other engines we have studied. Some outboard engines work on the four-stroke principle and others work on the two-stroke principle. An outboard engine has the same fuel, ignition, lubrication and cooling systems used in other small engines. The same troubleshooting, maintenance and tuneup procedures are used on outboards as on other small engines we have studied. Engine service operations are essentially the same as those used for other small engines. In this unit we will study the parts that are different in an outboard.

LET'S FIND OUT: When you finish reading and studying this unit, you should be able to:
1. Identify the main parts of an outboard engine.
2. List the steps used to remove an outboard engine carburetor.
3. Explain how to service an outboard engine carburetor.
4. List the steps used to install an outboard engine carburetor.

OUTBOARD ENGINE OPERATION

Most outboard boat engines like the one shown in Figure 23-1 are two-stroke engines. An outboard gets its name from the fact that it is mounted to the outside of the boat. The engine is divided into two general parts, as shown in Figure 23-2. The power head section is where the engine is mounted. The power from the engine is directed down through the lower unit to drive the propeller in the water. The operation of an outboard two-stroke engine is similiar to other two-strokes we have studied.

Figure 23-1. An outboard engine is a two-stroke engine. (Evinrude Motors)

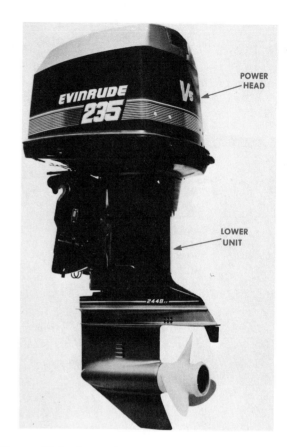

Figure 23-2. An outboard has a power head and a lower unit. (Evinrude Motors)

Figure 23-3. The power head is a two-cylinder engine.

ENGINE PARTS

Many outboard engines have more than one cylinder. They may have two, three, four, six or even more cylinders. The pistons in those cylinders are all connected to one common crankshaft.

The crankshaft is designed so that different cylinders have a power stroke at different times. This provides more than one power impulse to the crankshaft during each of its revolutions.

The power head shown in Figure 23-3 is a two-cylinder engine. An alternate firing order is used so that each cylinder delivers one power impulse per crankshaft rotation. The crankcase shown in Figure 23-4 houses the crankshaft. The crankcase may be divided in two pieces to remove the crankshaft. The cylinders are cast in one piece with the crankcase.

The piston, shown in Figure 23-5 with the piston rings, receives the force of combustion in the cylinder head, so it is necessary that both the

Figure 23-4. The crankcase for a two-cylinder outboard.

Figure 23-5. Piston and connecting rod from an outboard engine.

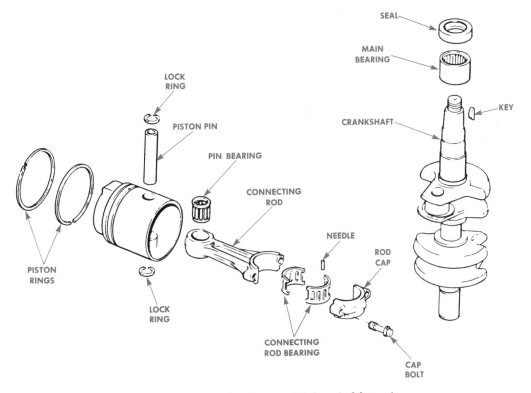

Figure 23-6. Exploded view of outboard power head parts. (Evinrude Motors)

pistons and piston rings be fitted properly to form a seal between the piston head and cylinder walls. To retain maximum power within the cylinder above the piston head, the cylinder must be perfectly round and the piston rings correctly seated in their grooves.

The connecting rods provide linkage between the piston and crankshaft. Connecting rod bearings include a roller bearing at the wrist pin, and split cage roller bearing at the crankshaft. The crankshaft is of the two-throw type and is supported by three main bearings. A single row roller bearing is at the upper journal. A split cage roller bearing at the center journal is aligned to the cylinder block by a dowel pin. A ball bearing at the bottom journal absorbs the radial and vertical thrust loads of the crankshaft. An exploded view of the piston and crankshaft assembly is shown in Figure 23-6. A sectional view of the entire engine and lower unit is shown in Figure 23-7.

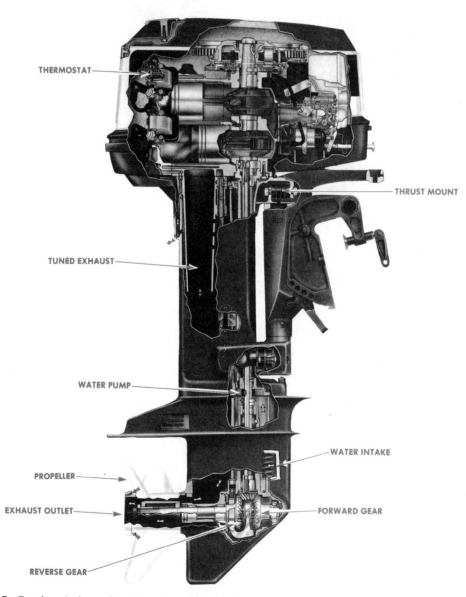

THERMOSTAT

THRUST MOUNT

TUNED EXHAUST

WATER PUMP

WATER INTAKE

PROPELLER

EXHAUST OUTLET

FORWARD GEAR

REVERSE GEAR

Figure 23-7. Sectional view of outboard engine. (Evinrude Motors)

OUTBOARD CARBURETOR SERVICE

An exploded view of an outboard carburetor similiar to that shown in Figure 23-8 makes a helpful guide in disassembling and reassembling the carburetor. Remove the strap and fuel hose from the carburetor. Remove the drain plug and drain the fuel bowl. Using a screwdriver, remove the low-speed needle valve, Figure 23-9. Using a special jet wrench, remove the high-speed orifice. Remove the float chamber, hinge pin, float and float valve assembly.

Clean all parts, except float and rubber parts, in solvent and blow dry. DO NOT dry parts with a

Figure 23-9. Location of low speed needle valve and drain plug.

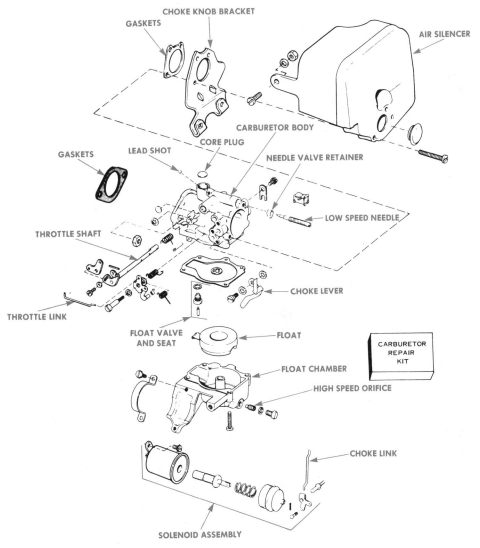

Figure 23-8. Exploded view of outboard carburetor. (Evinrude Motors)

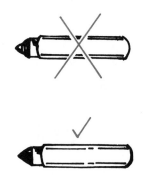

Figure 23-10. Check inlet float valve for wear.

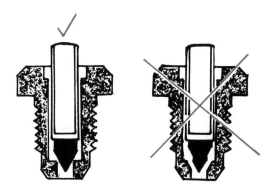

Figure 23-11. Inspect the inlet float valve seat for wear.

cloth as lint may cause trouble in the reassembled carburetor. Be sure all particles of gaskets are removed from gasket surfaces. Flush all passages in the carburetor body with solvent and remove any gummy deposits in a cold tank.

Inspect float and arm for wear or damage. Check float arm wear in the hinge pin and float valve contact areas. Replace if necessary. Inspect the float valve for grooves, nicks or scratches, Figure 23-10. If any are found, replace the float valve assembly. Gum or varnish on the float valve must be removed with solvent or in the cold tank. DO NOT attempt to alter the shape of the float valve. Check the float valve seat, Figure 23-11, with a magnifying glass; if the seat is nicked, scratched, or worn out-of-round, it will not give satisfactory service. The valve seat and valve are a matched set; if either is worn, both parts must be replaced. Use a new gasket when installing the valve seat.

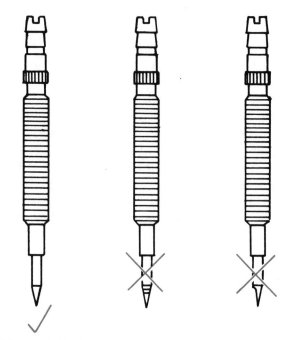

Figure 23-12. Inspect low-speed needle valve for wear.

Inspect the tapered end of the needle valve, Figure 23-12, for grooves, nicks, or scratches; replace if necessary. Wear on one side of the valve may indicate vibration from a damaged propeller. DO NOT attempt to alter the shape of the low-speed needle valve.

Clean out all the jets and passages, and the venturi, making sure no gum or varnish deposits remain. Dry after cleaning with compressed air. Check all gasket surfaces for nicks, scratches or distortion. Slight irregularities can be corrected with the use of a surface plate and emery cloth. Check throttle and choke shafts for excessive play. Check operation of the choke and throttle valves to be sure they correctly shut off air flow, yet move freely without binding. Replace the carburetor body if valves or shafts are excessively worn or damaged. Check hinge pin holes. Replace carburetor body if hinge pin holes are worn.

Reassemble the carburetor, paying particular attention to the following procedure: keep all dust, dirt and lint out of the carburetor during reassembly. Be sure that parts are clean and free from gum, varnish, and corrosion when reassembling them. Replace all gaskets and sealing washers. DO NOT attempt to use the original gaskets and washers, as leaks may develop after the engine is back in use.

Install the high-speed orifice and plug in the carburetor body. Replace the float valve seat and gasket, float valve, float and hinge pin. Check for correct positioning of the float. Turn the carburetor body upside down so the weight of the float closes the needle. The top of the float should be parallel with the rim of the casting. Reassemble the float chamber to the carburetor body, using a new gasket.

Install the low-speed needle and retainer. Use a screwdriver and turn the needle in until it gently touches its seat. Do not force the needle against the seat. Back off the needle 1¼ turns. This is a preliminary adjustment. A final adjustment must be made later with the engine running.

Place a new carburetor gasket on the intake manifold. Route the fuel hose to the fuel pump, and install the carburetor. Connect the cam follower link. Connect the fuel hose to the fuel pump using a new nylon hose strap. After the carburetor is installed, mount the engine in a test tank. Start and warm up the engine. Make carburetor adjustments specified by the engine manufacturer.

NEW TERMS

lower unit: The lower part of an outboard engine, used to transfer engine power to the propeller.

power head: The top part of an outboard engine where the power is developed.

SELF CHECK

1. List the two main parts of an outboard engine.
2. How many cylinders does an outboard engine have?
3. What is the function of an outboard engine power head?
4. What is the function of an outboard engine lower unit?
5. What are the main steps used in removing an outboard carburetor?
6. What kind of wear should you look for on a needle valve?
7. How should carburetor jets and passages be cleaned?
8. What are the main steps used to replace a carburetor on an outboard engine?

DISCUSSION TOPICS AND ACTIVITIES

1. Using an outboard engine, make a list of all the parts you can identify.
2. Using an owner's manual for an outboard engine, make a list of all the maintenance jobs that must be done to the engine.

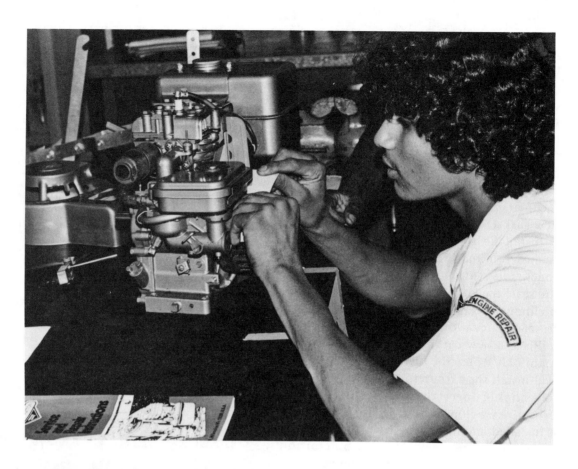

260

part 8
a career in small-engine service

Many career opportunities exist in the small-engine field. You might be involved in the making, selling or servicing of small engines. In making any kind of career choice, you must be sure that you are interested in and want to do the kind of work involved. The best way to find out is to try a career by working in a cooperative program, working part-time or during the summer. In this section we will try to give you an idea of what to expect.

261

unit 24

a career in small engines

It is not too early to think about a career. You should think about a career in each of the areas you study in school. Then, when the time comes for you to choose a career, or job, you can make the right choice. This unit will describe some of the careers in small engines. It will also give you an idea of what to expect working in the small-engines service field.

LET'S FIND OUT: **When you finish reading and studying this unit, you should be able to:**
1. **Describe the job of a small-engine mechanic.**
2. **Describe the job of a shop supervisor.**
3. **Explain what a parts specialist does.**
4. **List the duties of a small-engine shop owner.**
5. **Describe how to prepare for a career.**

CAREER AREAS

While big and complex, the small-engine industry may be divided into three general areas.
- Production: The turning of raw materials such as aluminum, steel and plastic into small engines.
- Sales: The selling of the manufactured engines.
- Service: The maintaining and repairing of small engines.

There are, of course, many job opportunities in all three areas of the small-engine industry. In this book, we are most interested in those having to do with service.

MECHANIC/TECHNICIAN

Small-engine mechanics or technicians, Figure 24-1, perform preventive maintenance, diagnose, disassemble and make necessary repairs on small engines. Preventive maintenance is the periodic examination, and the adjustment, repair or replacement of parts before they wear out. Troubleshooting checklists are used to make sure there is a complete examination of all systems. When mechanical or electrical problems exist, a description of the difficulty is obtained from the operator, the problem is trouble-shot as to probable cause, and adjustments, repairs or replacements are made until the problem is solved.

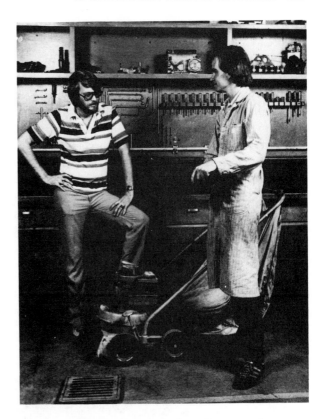

Figure 24-2. The service writer discusses the customer's problems.

Figure 24-1. A small-engine mechanic repairs small engines.

SUPERVISOR

Many mechanics eventually work hard enough to advance into a supervisory position. These jobs include shop supervisor, service writer and service manager. The shop supervisor is in charge of a number of mechanics in a service department. Supervisors must schedule the work and make sure it is done correctly. The service writer, Figure 24-2, greets the customer and discusses the customer's problems. The service writer then pre-

pares the cost estimates the billings required to get the needed service performed. The service manager is responsible for all the different service departments in a large garage or dealership.

PARTS SPECIALIST

In order to repair a small engine correctly, a mechanic must have the correct replacement parts. The ordering, cataloging, storing and selling of these parts is the responsibility of the parts specialist. This worker must have a thorough knowledge of parts and their interchangeability as well as special training in reading catalogs and making inventories. Small-engine parts suppliers employ personnel to inventory, order and sell parts. In addition, many suppliers operate machine shops and do component rebuilding. These services require the employment of small-engine mechanics.

OWNERSHIP

Many small-engine mechanics eventually open their own repair shops, Figure 24-3. The shop may specialize in one small-engine service area such as lawn equipment, chain saws, outboards or motorcycles. Usually the shop sells new products and parts as well as offering repair service.

Owning a shop requires not only a background in small engines but a knowledge of business practice. An owner must know how to keep sales and service records, promote sales, make out a payroll and provide benefits for employees as well as maintain an efficient shop.

A DAY ON THE JOB

What's it like to work in a small-engine repair shop? Let's look in on JOE LEROY'S LAWN-MOWER SHOP. Joe owns his own shop and generally works alone except during the busy summer months when he hires a helper or two. It's nine o'clock and Joe is heading toward his front door to open up. A customer is waiting for Joe at the front door with a lawnmower.

"Good morning, Mr. Smith."

"Good morning, Joe. I have been pulling on this darn lawnmower rope half the morning and I can't get it to start."

Figure 24-3. *Top,* many small engine mechanics own their own shops. *Bottom,* a modern small-engine repair facility.

"Push it on into the shop, Mr. Smith, and I'll have a look."

Joe turns on the lights in the shop and moves Mr. Smith's lawnmower to an area near his tools. A quick look at the outside of the engine satisfies Joe that the magneto switch is on and the spark plug wire is connected properly. Joe removes the fuel-tank cap and looks inside the tank. There is plenty of fuel but it smells stale.

"Mr. Smith, when is the last time you started this engine?" he asks.

"Not for several weeks; we've been on vacation."

Joe gets a wrench from his tool panel and removes the spark plug. As he suspected, the plug is dry. "This carburetor is plugged up with stale fuel. It will need to be cleaned."

"O.K., Joe, but can I get it back this week?"

"Sure. Let me give you a job estimate."

Joe gets out a pad of job estimate forms from his desk. The forms are used to estimate how much a repair job will cost the customer. The form is a kind of contract. If the charges of actually repairing Mr. Smith's lawnmower turn out to be more than the estimate, Joe knows he much check with Mr. Smith before doing the work. Joe knows that to remove, clean and rebuild this kind of carburetor will take him about an hour. He charges $10 per hour. This is called a labor charge. The parts and gaskets for the job sell for $8. Joe writes the labor and parts charges on the estimate, Figure 24-4.

"Here's the estimate, Mr. Smith. It should come to $18. If I find anything else wrong, I'll call

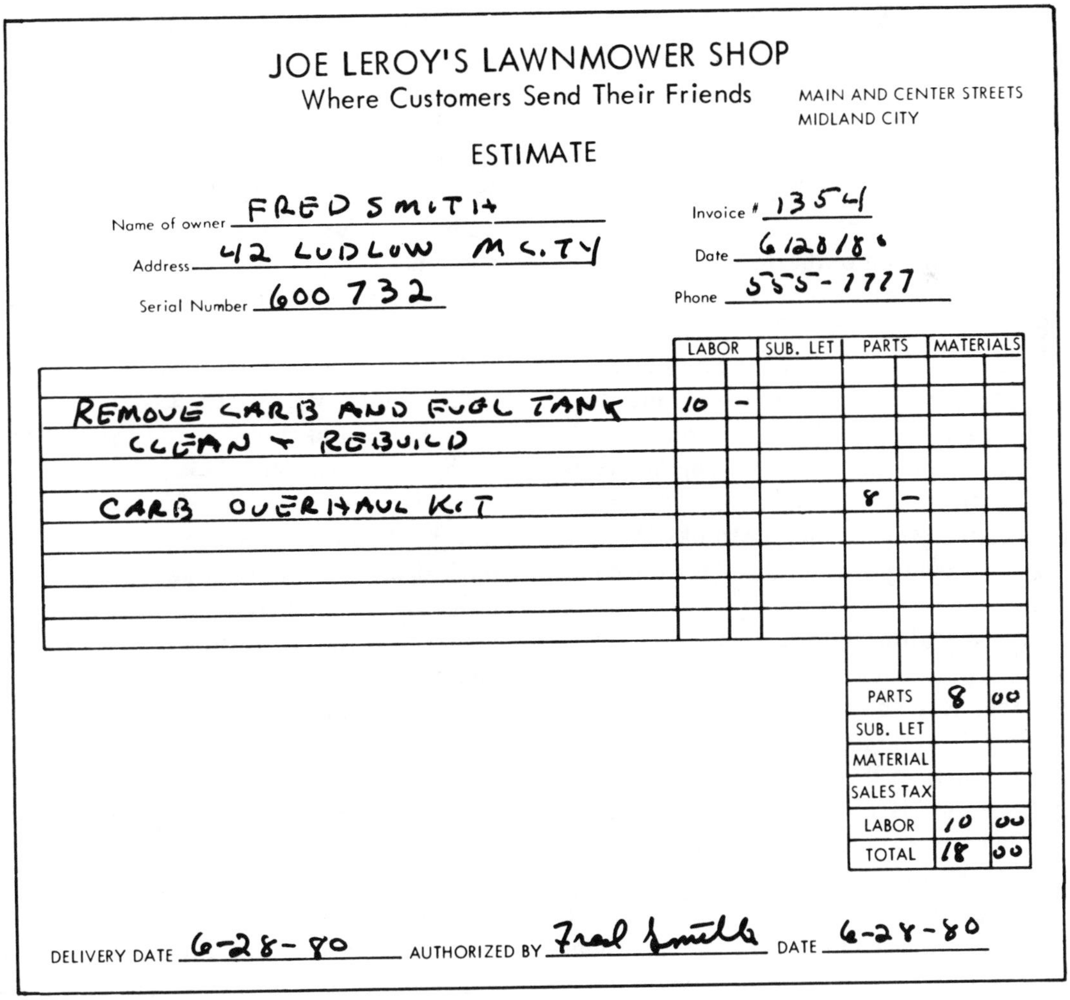

Figure 24-4. A sample estimate.

you. If this is O.K. with you, sign on the bottom. I should have it ready by this evening."

When Mr. Smith leaves, Joe begins work on some of the other jobs that were promised for today. A lawnmower blade to be sharpened for Homer over at the bank. An engine overhaul that has to be finished by Friday. As Joe is putting the finishing touches on the engine, the mail carrier delivers a package.

"Hi, Joe, this is your package," she says.

"Good. I hope these are the gasket sets I ordered."

The package opened, Joe finds that they are the gasket sets he needs for engine overhauls. He must put the gasket sets on his parts shelves. Parts are placed on the shelf by part number. Joe keeps a running record of the parts he has. This record is called an inventory. As Joe uses parts from his inventory, he must order a new supply. Joe buys his parts wholesale. This means he buys parts at prices available only to shop owners. He sells the parts at a markup called *list price*. This way Joe makes a profit on both parts and labor.

In the mail with the parts, Joe also finds several bills. These are for his rent, electricity and gas. Joe has bills like this every month. He totals these bills along with his cost for parts and the wages he pays his helpers when they work. This is Joe's overhead, or what it costs him to operate his business. Joe figures his profit by subtracting his monthly overhead from his income for the month.

Joe works some more on the engine he is rebuilding until he notices the time. "Better get Mr. Smith's lawnmower done," he thinks. As he is removing the carburetor, another customer enters the shop.

"Hello, Mrs. Maloney, what can I do for you?" asks Joe.

"Hello Joe, this spark plug from my lawnmower is bad. I would like to buy a new one. Do you have one in stock?"

Joe pulls a new spark plug of the same number out of his inventory and sells it to Mrs. Maloney. He gets back to work on the carburetor.

Joe disassembles the carburetor and places the parts in his carburetor tank. While he lets them soak, he works on his parts orders. Several parts must be reordered to keep his inventory up to date. Joe knows that if he doesn't have a part he needs a customer's engine may be held up from delivery. This may cost him some of his repeat business.

In about an hour, the carburetor in the tank is cleaned. Joe removes the carburetor and flushes it with hot water. He dries the parts with air. New parts are pulled from inventory and installed in the carburetor.

The rebuilt carburetor is installed on the engine. Joe fills the fuel tank with fresh fuel. He makes preliminary carburetor adjustments and double-checks the engine. Satisfied that it is ready to go, Joe gives the engine a pull of the rope. In two pulls the engine catches and runs. It sounds good. Joe lets it warm up and makes final fuel-mixture adjustments.

"That should do it," he thinks. Joe installs the air cleaner and pushes the mower to a side of his shop reserved for completed jobs.

"Time to make out the bill for this job." Joe gets out the bill pad. A bill or invoice, Figure 24-5, lists the parts and labor actually used on the job. While very similar to the form used for an estimate, it is used for the actual charge to the customer.

As Joe was completing the bill, he looked up to see Mr. Smith coming in the door.

"Hi, Joe, did you get it done?"

"Sure did, Mr. Smith. Listen to it run." Joe gave one pull of the rope and the engine fired.

"Sounds great, Joe."

Joe turned the engine off. "Here's the bill, Mr. Smith." It was $18. Mr. Smith paid the bill and pushed the lawnmower out of the shop.

Joe filed the bill and put the money in his cash register. He looked at the clock. It was six o'clock and time to quit. Joe locked up the shop and headed home for dinner.

PREPARING FOR A CAREER

There are many things you can do to get ready for a career. You should try to learn all you can about careers. Talk to people that work in areas in which you are interested. Find out the good and bad things about their jobs. There are books on

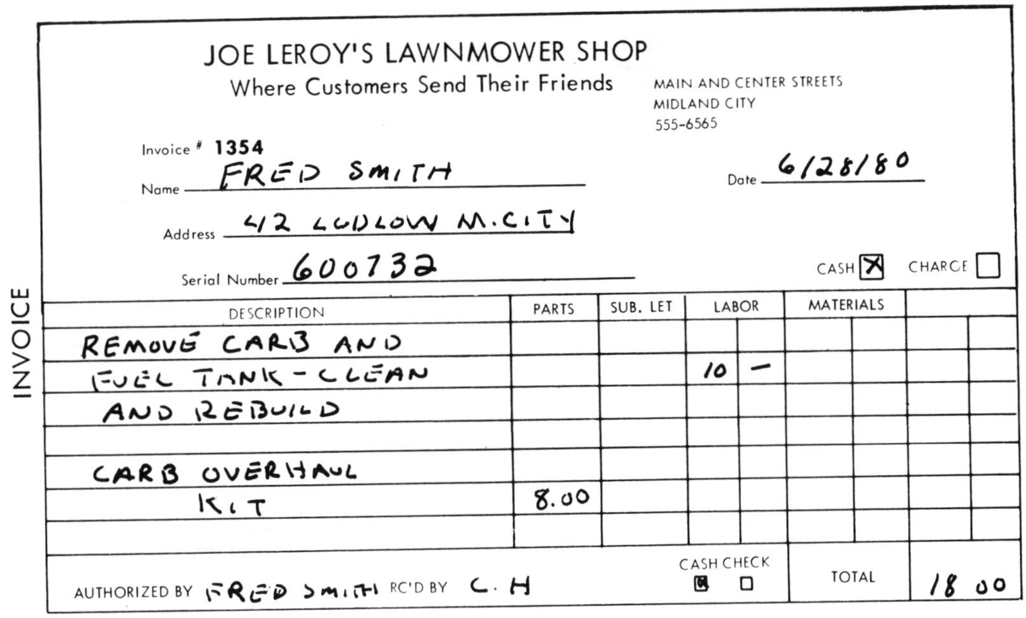

JOE LEROY'S LAWNMOWER SHOP
Where Customers Send Their Friends

MAIN AND CENTER STREETS
MIDLAND CITY
555-6565

Invoice # **1354**

Name _FRED SMITH_

Date _6/28/80_

Address _4/2 LUDLOW M.CITY_

Serial Number _600732_

CASH ☒   CHARGE ☐

DESCRIPTION	PARTS	SUB. LET	LABOR	MATERIALS	
REMOVE CARB AND					
FUEL TANK - CLEAN			10 —		
AND REBUILD					
CARB OVERHAUL					
KIT	8.00				

AUTHORIZED BY _FRED SMITH_ RC'D BY _C. H_

CASH ☒ CHECK ☐

TOTAL _18 00_

Figure 24-5. A sample bill or invoice.

many careers. Look through these books in your school or local library.

There are many people who can help you find out about a career. Your small-engines teacher can help. You school counselor knows about careers.

Different careers require different skills and interests. Before you make a career choice, you should find out whether you have the right skills and interests. Your teachers and counselors can help you find out what your skills and interests are.

NEW TERMS

career: A job or occupation.

bill: A listing of labor and parts charges for a repair job.

estimate: A list of estimated parts and labor charges for a repair job.

inventory: A list of the parts on hand.

invoice: The same as a bill.

mechanic: Worker who services a machine such as a small engine.

overhead: Money spent on items necessary to operate a business.

technician: Same as mechanic.

SELF CHECK

1. What is a career?
2. What do workers in the manufacturing area do?
3. What do workers in the sales area do?
4. What do workers in the sevice area do?
5. What is a mechanic?
6. Describe what a mechanic does.
7. What does a parts specialist do?
8. What does a supervisor do?
9. What are some of the duties of an owner of a small-engine shop?
10. List the things you can do to get ready for a career.

DISCUSSION TOPICS AND ACTIVITIES

1. Information on occupations can be found in two reference books called OCCUPATIONAL HANDBOOK and DICTIONARY OF OCCUPATIONAL TITLES. Locate these books in the library and research one of the jobs listed.
2. Visit a local small-engine service business and report on what you see.

appendix

4ths	8ths	16ths	32nds	64ths	to 2 places	to 3 places	to 4 places	mm	4ths	8ths	16ths	32nds	64ths	to 2 places	to 3 places	to 4 places	mm
				1/64	**0.02**	0.016	0.0156	0.3969					33/64	**0.52**	0.516	0.5156	13.0968
			1/32		**0.03**	0.031	0.0312	0.7937				17/32		**0.53**	0.531	0.5312	13.4937
				3/64	**0.05**	0.047	0.0469	1.1906					35/64	**0.55**	0.547	0.5469	13.8906
		1/16			**0.06**	0.062	0.0625	1.5875			9/16			**0.56**	0.562	0.5625	14.2875
				5/64	**0.08**	0.078	0.0781	1.9844					37/64	**0.58**	0.578	0.5781	14.6844
			3/32		**0.09**	0.094	0.0938	2.3812				19/32		**0.59**	0.594	0.5938	15.0812
				7/64	**0.11**	0.109	0.1094	2.7781					39/64	**0.61**	0.609	0.6094	15.4781
	1/8				**0.12**	0.125	0.1250	3.1750		5/8				**0.62**	0.625	0.6250	15.8750
				9/64	**0.14**	0.141	0.1406	3.5719					41/64	**0.64**	0.641	0.6406	16.2719
			5/32		**0.16**	0.156	0.1562	3.9687				21/32		**0.66**	0.656	0.6562	16.6687
				11/64	**0.17**	0.172	0.1719	4.3656					43/64	**0.67**	0.672	0.6719	17.0656
		3/16			**0.19**	0.188	0.1875	47625			11/16			**0.69**	0.688	0.6875	17.4625
				13/64	**0.20**	0.203	0.2031	5.1594					45/64	**0.70**	0.703	0.7031	17.8594
			7/32		**0.22**	0.219	0.2188	5.5562				23/32		**0.72**	0.719	0.7188	18.2562
				15/64	**0.23**	0.234	0.2344	5.9531					47/64	**0.73**	0.734	0.7344	18.6532
1/4					**0.25**	0.250	0.2500	6.3500	3/4					**0.75**	0.750	0.7500	19.0500
				17/64	**0.27**	0.266	0.2656	6.7469					49/64	**0.77**	0.766	0.7656	19.4469
			9/32		**0.28**	0.281	0.2812	7.1437				25/32		**0.78**	0.781	0.7812	19.8433
				19/64	**0.30**	0.297	0.2969	7.5406					51/64	**0.80**	0.797	0.7969	20.2402
		5/16			**0.31**	0.312	0.3125	7.9375			13/16			**0.81**	0.812	0.8125	20.6375
				21/64	**0.33**	0.328	0.3281	8.3344					53/64	**0.83**	0.828	0.8281	21.0344
			11/32		**0.34**	0.344	0.3438	8.7312				27/32		**0.84**	0.844	0.8438	21.4312
				23/64	**0.36**	0.359	0.3594	9.1281					55/64	**0.86**	0.859	0.8594	21.8281
	3/8				**0.38**	0.375	0.3750	9.5250		7/8				**0.88**	0.875	0.8750	22.2250
				25/64	**0.39**	0.391	0.3906	9.9219					57/64	**0.89**	0.891	0.8906	22.6219
			13/32		**0.41**	0.406	0.4062	10.3187				29/32		**0.91**	0.906	0.9062	23.0187
				27/64	**0.42**	0.422	0.4219	10.7156					59/64	**0.92**	0.922	0.9219	23.4156
		7/16			**0.44**	0.438	0.4375	11.1125			15/16			**0.94**	0.938	0.9375	23.8125
				29/64	**0.45**	0.453	0.4531	11.5094					61/64	**0.95**	0.953	0.9531	24.2094
			15/32		**0.47**	0.469	0.4688	11.9062				31/32		**0.97**	0.969	0.9688	24.6062
				31/64	**0.48**	0.484	0.4844	12.3031					63/64	**0.98**	0.984	0.9844	25.0031
1/2					**0.50**	0.500	0.5000	12.7000	1					**1.00**	1.000	1.0000	25.4000

Fractional, Number, and Letter Sizes for Twist Drills

DRILL NO.	FRAC.	DECI.	DRILL NO.	FRAC.	DECI.	DRILL NO.	FRAC	DECI.	DRILL NO.	FRAC.	DECI.	DRILL NO.	FRAC.	DECI.	DRILL NO.	FRAC.	DECI.	DRILL NO.	FRAC.	DECI.	DRILL NO.	FRAC.	DECI.
80	--	.0135	60	--	.0400	--	3/32	.0938	24	--	.152	6	--	.204	L	--	.290	--	13/32	.406	--	47/64	.734
79	--	.0145	59	--	.0410				23	--	.154	5	--	.206				Z	--	.413	--	3/4	.750
--	1/64	.0156	58	--	.0420	41	--	.0960	--	5/32	.156	4	--	.209	M	--	.295	--	27/64	.422	--	49/64	.766
78	--	.0160				40	--	.0980	22	--	.157				--	19/64	.297	--	7/16	.438			
77	--	.0180	57	--	.0430	39	--	.0995	21	--	.159	3	--	.213	N	--	.302	--	29/64	.453	--	25/32	.781
			56	--	.0465	38	--	.1015				--	7/32	.219							--	51/64	.797
76	--	.0200	--	3/64	.0469	37	--	.1040	20	--	.161	2	--	.221	O	--	.316	--	15/32	.469	--	13/16	.813
75	--	.0210	55	--	.0520				19	--	.166	1	--	.228				--	31/64	.484	--	53/64	.828
74	--	.0225	54	--	.0550	36	--	.1065	18	--	.170	A	--	.234	P	--	.323	--	1/2	.500	--	27/32	.844
73	--	.0240				--	7/64	.1094	--	11/64	.172							--	33/64	.516			
72	--	.0250	53	--	.0595	35	--	.1100	17	--	.173	--	15/64	.234	Q	--	.332	--	17/32	.531	--	55/64	.859
			--	1/16	.0625	34	--	.1110				B	--	.238	R	--	.339				--	7/8	.875
71	--	.0260	52	--	.0635	33	--	.1130	16	--	.177	C	--	.242	--	11/32	.344	--	35/64	.547	--	57/64	.891
70	--	.0280	51	--	.0670				15	--	.180	D	--	.246				--	9/16	.562	--	29/32	.906
69	--	.0292	50	--	.0700	32	--	.116	14	--	.182	--	1/4	.250				--	37/64	.578	--	59/64	.922
68	--	.0310				31	--	.120	13	--	.185				S	--	.348	--	19/32	.594			
--	1/32	.0313	49	--	.0730	--	1/8	.125				E	--	.250	T	--	.358	--	39/64	.609	--	15/16	.938
			48	--	.0760	30	--	.129	--	3/16	.188	F	--	.257	U	--	.368				--	61/64	.953
67	--	.0320	--	5/64	.0781	29	--	.136				G	--	.261	--	3/8	.375	--	5/8	.625	--	31/32	.969
66	--	.0330	47	--	.0785				12	--	.189	--	17/64	.266				--	41/64	.641	--	63/64	.984
65	--	.0350	46	--	.0810				11	--	.191	H	--	.266				--	21/32	.656	--	1	1.000
64	--	.0360				--	9/64	.140	10	--	.194				V	--	.377	--	43/64	.672			
63	--	.0370	45	--	.0820	28	--	.141	9	--	.196	I	--	.272	W	--	.386	--	11/16	.688			
			44	--	.0860	27	--	.144	8	--	.199	J	--	.277	--	25/64	.391						
62	--	.0380	43	--	.0890	26	--	.147	7	--	.201	--	9/32	.281	X	--	.397	--	45/64	.703			
61	--	.0390	42	--	.0935	25	--	.150	--	13/64	.203	K	--	.281	Y	--	.404	--	23/32	.719			

basic terms

A

adjustable wrench: A wrench designed to adjust to different sizes of bolts and nuts.

air cleaner: A filter mounted on the carburetor to clean the air before it enters the carburetor.

air cooling: Cooling engine parts by circulating air around them.

air impact wrench: A wrench powered by compressed air.

air pump: Pump used with an air cooling system to force air around hot parts.

apex seals: Seal on the ends or apex (tip) of the rotary engine rotor used to close off the firing chamber.

armature gap: The space between the armature and flywheel magnets.

atom: Small particle of matter.

B

bill: A listing of labor and parts charges for a repair job. Sometimes called an *invoice*.

bolt: A threaded fastener used with a nut to hold small-engine parts together.

bore: The diameter of the cylinder.

boring: Machining away metal from a worn cylinder with a boring bar for the installation of new oversize pistons.

brake horsepower: Horsepower measured at the engine's flywheel. Abbreviated BHP.

brake mean effective pressure: The average pressure exerted on the piston during one operating cycle. Abbreviated BMEP.

breaker point gap: The space between the movable and stationary ignition points when they are open.

breaker points: The switch used in the ignition primary system to control coil operation.

break-in: The wearing-in to improve the fit between the piston rings and cylinder walls.

bypass port: Passageway between the crankcase and combustion area in a two-stroke engine. Sometimes called a transfer port.

C

camshaft: A shaft with lobes used to open the engine's valves at the proper time.

capacitive discharge ignition system: An ignition system that uses the energy stored in a capacitor to develop high voltage.

capacitor: An electrical device used to store or soak up a surge of electricity.

carburetor: A part that mixes air and fuel in the correct amounts to burn in the engine.

carburetor service: The removal, disassembly, cleaning and reassembly of a carburetor.

career: A job or occupation.

chisel: A bar of hardened steel with a cutting edge ground on one end. It is driven with a hammer to cut metal.

circuit: A complete path for electrical current flow.

cleaning equipment: Equipment used to clean small-engine parts.

coil: An electrical device used to step up voltage for ignition.

cold tank cleaner: A tank with a cold solution for cleaning nonferrous metal parts such as aluminum.

combination wrench: A wrench with a box end at one end and an open end at the other.

combustion chamber: Part of the engine in which the burning of the air and fuel takes place.

compressed air: Air under pressure.

compression gage: A tester used to determine whether an engine has enough compression.

compression ratio: The space into which the air-fuel mixture is squeezed during the compression stroke, compared to its original volume.

compression release: A valve in the cylinder head that allows pressure in the cylinder to be released.

compression ring: A piston ring used to seal compression pressures in the combustion chamber.

compression stroke: The stroke of the four-stroke engine in which the air-fuel mixture is compressed.

condenser: The capacitor used in the ignition primary to prevent contact breaker point arcing.

conductor: A material that allows electrical current flow.

connecting rod: An engine part that connects the piston to the crankshaft.

connecting rod bearing: The bearing used between the connecting rod and the crankshaft.

connecting rod bearing clearance: The space for oil between the crankshaft and the connecting rod bearing.

coolant: Liquid used in liquid-cooling system to carry away heat; usually a mixture of ethylene glycol and water.

coolant pump: Pump used to circulate coolant around hot engine parts.

cooling fin: Metal fin used on air-cooled engine parts to move heat away from the parts.

cooling system: An engine system designed to circulate air or coolant around engine parts to prevent excessive heat build-up and parts damage.

counterbored ring: A ring with a section removed to allow twisting and improvement of sealing.

crankcase: The part of the engine that supports the crankshaft.

crankcase breather: A valve assembly used to vent crankcase pressure.

crankshaft: An offset shaft to which the piston and connecting rods are attached.

crankshaft end-play: Movement of the crankshaft measured with a dial indicator.

crankshaft out-of-round: A wear condition in which a crankshaft journal is no longer round.

crankshaft taper: A wear condition in which one end of a crankshaft journal is larger than another.

current: The flow of electrons in an electrical circuit. Measured in amperes and abbreviated *I*.

customary measuring system: One of two main measuring systems in use in the world. Most common system used in the United States.

customary system units: Units of measure based on the yard. The yard is divided into feet and inches.

cylinder: A hole in which an engine piston rides.

cylinder gage: A dial gage made to measure cylinder taper.

cylinder head: Large casting bolted to the top of the engine containing the combustion chamber and valves.

cylinder head warpage: A condition in which the cylinder head is not flat.

cylinder out-of-round: A wear condition in which the cylinder is worn more in one direction than another.

cylinder taper: A wear condition in which the top of the cylinder is larger than the bottom.

D

dial indicator: A gage used to measure movement or "play" and contour or "runout" of a small-engine part.

diaphragm: A piece of flexible material stretched across the pressure chamber of a carburetor or fuel pump used to control the amount of fuel that enters the chamber.

diaphragm carburetor: A carburetor that controls with a diaphragm the amount of fuel admitted.

die: A tool used to cut external threads.

diesel engine: An engine that uses the heat of compression to ignite the air-fuel mixture.

displacement: The volume swept or displaced by the pistons of an engine.

dowel pin: A round metal pin that fits into drilled holes to position two mating parts.

dry paper air cleaner: An air cleaner that uses a paper element to filter air.

dry sump lubrication: A lubrication system that uses a scavenge pump to pull oil out of the engine into an oil tank.

dynamometer: Equipment used to measure torque and calculate horsepower.

E

efficiency: How well a machine such as an engine converts energy into useful work.

electric drill: A drill powered by electricity.

electricity: The flow of electrons from one atom to another.

electric wrench: A wrench powered by electricity.

energy: The ability to do work.

engine: A device that converts heat energy in a fuel into mechanical energy that can be used to work such as power a lawn mower.

estimate: A list of probable parts and labor charges for a repair job.

exhaust port: Passage used to route out burned gases from the cylinder.

exhaust stroke: The stroke of a four-stroke-cycle engine during which burned gases are expelled.

exhaust valve: Valve used to control the flow of burned exhaust gases from the cylinder.

expander: A flexible spring placed behind a piston ring to increase spring tension.

F

feeler gage: A measuring tool used to measure accurately the space between two surfaces.

file: A hardened steel tool with rows of cutting edges used to remove metal for polishing, smoothing or shaping.

fire extinguisher: A pressurized container of foam or dry chemicals used to extinguish a fire.

fire prevention: Keeping fires from starting by following safe practices.

firing chamber: The area in the rotary engine where power is developed.

fit: The clearance between parts.

float: A part of the carburetor that regulates the correct amount of fuel.

float adjustment: Setting the float level in the carburetor by bending the float tang.

float carburetor: A carburetor that controls the amount of fuel admitted with a float.

float-level gage: A tool used to measure the float level in a sliding-valve carburetor.

flywheel: Heavy wheel used to store energy and smooth out engine operation.

force: A push or a pull.

four-stroke-cycle engine: Engine that develops power using four strokes of a piston.

friction: Resistance to motion between two parts that causes wear and heat.

G

gasket: A soft material used between two engine parts to form a pressure seal.

glaze-breaking: Roughing up the surface of a cylinder with an abrasive stone to help the rings seat.

guide bushing: Tool used to guide a reamer when reconditioning main bearings.

H

hacksaw: A saw for cutting metal.

hammer: A tool used to drive or pound on an object. Hammers for automotive use may have a hard or a soft head.

hand tool: Tool that uses no power except hand power.

hex wrench: A wrench used to tighten or loosen allen or hollow-head screws.

high-speed adjustment: The setting of a carburetor's air-fuel mixture at high speed.

high-speed adjustment screw: A screw on the carburetor used to regulate the fuel mixture at high speed.

honing: Removing metal from a cylinder with abrasive stones to straighten it.

horsepower: Term used to describe the power developed by an engine. One horsepower equals 33,000 foot-pounds of work per minute.

horsepower rating: Different ways in which horsepower is measured and specified.

I

idle-speed adjustment: The setting of a carburetor's idle speed or RPM.

ignition cable: High-voltage ignition wire used to carry secondary voltage.

ignition system: The electrical system that provides the high-voltage spark to ignite the air-fuel mixture in the cylinder.

indicated horsepower: A laboratory horsepower measurement based upon the power developed in the engine's cylinders.

induction: The transfer of energy from one object to another without the objects touching.

inside micrometer: A measuring tool used to measure the size of holes such as a small-engine cylinder.

intake port: Passage in the cylinder head used to route the flow of air and fuel into the cylinder.

intake stroke: The stroke of the four-stroke-cycle engine in which air and fuel enter the engine.

intake valve: Valve used to control the flow of air and fuel into the engine.

insulator: A material that prevents the flow of electricity.

internal combustion engine: An engine such as the four-stroke-cycle engine in which the burning of the fuel takes place inside the engine.

inventory: A list of the parts on hand.

invoice: The same as a bill.

K

key: A small hardened piece of metal used with a gear or pulley to lock it to a shaft.

knock: A metallic noise in an engine that generally indicates a problem.

L

linear horsepower: Horsepower used to pull in a straight line.

lobe: A raised bump on the camshaft used to lift the valve.

lobe height: The distance between the heel and the top of a cam lobe.

loop scavenge: A type of two-stroke engine in which the air-fuel mixture loops through holes in the piston skirt.

lower end: The crankshaft area of a moped or motorcycle engine.

lower unit: The lower part of an outboard engine used to transfer engine power to the propeller.

low-speed adjustment: The setting of a carburetor's air-fuel mixture at low speed.

low-speed adjustment screw: Carburetor screw used to regulate mixture at low speed.

lubrication: Reducing of friction in an engine by providing oil between moving parts.

M

magneto: Device used to develop the high voltage necessary for ignition.

margin: The outside rim of the valve that gets thinner as the face is ground.

mechanic: Worker who services a product such as a small engine.

metalworking tools: Tools used to cut or shape metal.

metric measuring system: One of the two main measuring systems in use in the world. Used for a long time in other countries, it is being adopted in the United States.

metric units: Units of measure based upon a meter and decimal steps of the meter.

multi-piece crankshaft: A crankshaft that may be disassembled to replace a journal.

N

narrowing: Removing part of the valve seat to make it narrow for better valve fit.

nonthreaded fastener: device to hold small-engine components together without the use of threads.

nut: A small fastener having internal threads used with bolts and screws.

O

oil: A fluid used for lubrication.

oil bath air cleaner: An air cleaner that uses a container of oil to trap dirt going into the engine.

oil foam air cleaner: An air cleaner that uses a sponge-like foam element to trap dirt.

oil level: The amount of oil inside an engine.

open-end wrench: Wrench with an opening at the end to allow it to be positioned on a bolt or nut.

output shaft: A shaft in a rotary engine that is driven by the rotor. The power to drive the vehicle's wheels is taken off this shaft.

outside micrometer: A measuring tool used to measure the outside of an object such as a crankshaft or piston.

overhead: Money spent on items necessary to operate a business.

P

phillips screwdriver: A screwdriver with a point on the blade or tip used for driving phillips head screws.

piston: Round metal part attached to the connecting rod which slides up and down in the cylinder.

piston pin: Pin used to attach the piston to the connecting rod.

piston ring: Expanding sealing ring placed in a groove around the piston.

piston ring end-gap: The space between the ends of a piston ring when it is in the cylinder.

piston ring side clearance: The space between a piston ring and ring land.

piston ring tool: Tool used to expand piston rings for removal.

plastic string: A string used to determine bearing clearance.

pliers: Tool designed for gripping objects that wrenches or screwdrivers will not fit.

power: The speed at which work is done.

power head: The top part of an outboard engine where the power is developed.

power stroke: The stroke of the four-stroke-cycle engine in which power is delivered to the crankshaft.

power tool: Tool powered by electricity, compressed air or hydraulic fluid. Power tools must be used carefully to prevent injury.

premix lubrication: Oil and gasoline mixed together for lubrication.

pressure lubrication: A system that uses a pump to force oil into engine parts.

R

radiator: A large heat exchanger located in front of the engine.

reed valve: A leaf-shaped valve used to control the flow of air-fuel mixture into the crankcase.

rotary engine: An engine with an elliptical firing chamber and a triangular shaped rotor that develops power in a rotary motion. Same as a Wankel engine.

rotary horsepower: Horsepower developed in a rotary motion such as by an engine's crankshaft.

rotary valve: A crankshaft-mounted valve that controls air-fuel mixture flow into a motorcycle engine.

rotor: The triangle-shaped part in a rotary engine that has the same job as a piston in a four-stroke-cycle engine.

rotor seals: Apex, corner and side seals used to prevent leakage out of the firing chamber.

rule: The simplest of all measuring tools. It is a flat length of wood, plastic or metal divided into a number of measuring units.

S

saddle bore: The large hole in the bottom of the connecting rod that fits around the crankshaft.

screw: A threaded fastener that fits into a threaded hole.

screwdriver: Tool used to install or remove screws.

seal: A device with a flexible lip used to keep fluids inside an area.

service classification: A rating system used for engine oil in which letters are used to describe the degree to which the oil is able to stand up in service. The classification SA is the lowest and SE the highest.

side seal: Seal installed in the side of a rotor to close off the firing chamber.

sliding-valve carburetor: A carburetor that uses a sliding valve attached to a needle to regulate fuel.

slotted tip screwdriver: A screwdriver with a tapered blade or tip made to drive common slotted screws.

small-hole gage: A measuring tool consisting of a split sphere and an internal wedge used to measure the inside of small holes such as valve guides.

socket handle and attachments: Tool used to drive a socket wrench.

socket wrench: A tool that fits all the way around a bolt or nut and is made to be detached from a handle.

solvent cleaner: A tank in which cleaning solvent is used to wash off oil and grease from small-engine parts.

spark plug: Ignition system part used to create a spark in the combustion chamber.

spark plug analysis chart: A chart showing common problems of spark plugs.

spark plug gap: The space between the center and ground electrode of a spark plug.

spark plug socket: A special socket made to remove spark plugs without damaging them.

splash lubrication: A system in which oil is splashed on engine parts for lubrication.

stroke: The movement of the piston in the cylinder, controlled and measured by the offset of the crankshaft.

stud: A fastener with threads at both ends.

T

tachometer: An electronic instrument used to measure engine speed or RPM.

tap: A tool used to cut internal threads.

technician: Same as mechanic.

telescoping gage: A measuring tool with a spring-loaded rod, the parts of which slide together in a cylinder. It is used to measure the inside of a hole.

thermal efficiency: How well an engine changes the chemical energy in gasoline to heat energy.

thermostat: A device in the cooling system used to control the flow of coolant.

thread designation: The system used to indicate the size of the threads on threaded fasteners.

threaded fasteners: Fasteners that use threads to hold two components together.

thin-wall guide: A thin bushing used to recondition a worn valve guide.

torque: A turning or twisting effort or force.

torque wrench: A wrench designed to tighten bolts or nuts to the correct tightness or torque.

transfer port: Same as bypass port.

troubleshooting: The logical step-by-step procedure used to locate an engine problem.

troubleshooting chart: A chart listing problems along with causes and cures.

tune-up: The replacing of worn ignition parts and adjustment of other systems to get the best performance from an engine.

twist drill: A hardened cutting tool made to cut or drill a hole.

two-stroke-cycle engine: An engine that develops power in two piston strokes or one crankshaft revolution.

U

upper end: The cylinder area of a moped or motorcycle engine.

V

valve: Device for opening and closing a port.

valve grinding: Machining the valve face by grinding.

valve guide: A part installed to support and maintain alignment of the valve.

valve guide clearance: The space between the valve guide and valve stem.

valve lash adjustment: The setting of the valve clearance or lash to the correct specifications.

valve lifter: Component that rides on the cam and pushes open the valve.

valve seat cutting: Reconditioning a valve seat by cutting with a hardened steel cutting tool.

valve seat grinding: Reconditioning a valve seat by resurfacing with a grinding stone.

valve seat lapping: Using an abrasive compound to remove metal from a valve seat and face.

valve spring: Coil spring used to close a valve.

valve spring compressor: Tool used to push together a valve spring so the retainer can be removed.

valve spring tension: The pressure of a valve spring measured when the spring is compressed.

valve train: An assembly of parts in an engine that opens and closes the passageways for the intake of air and fuel as well as the exhaust of burned gases.

viscosity: The thickness or thinness of the oil.

voltage: The source of potential energy in an electrical system. Measured in volts and abbreviated *E*.

volumetric efficiency: The ratio of an engine's cylinder volume to the volume filled by air and fuel during engine operation.

W

Wankel engine: Same as rotary engine. Named after the inventor, Felix Wankel.

washer: Fastener used with bolts, screws, studs and nuts to prevent them from vibrating loose and to distribute the clamping force.

water jacket: Passage in the cylinder block and head for coolant flow.

work: What is done when a force travels through a distance.

index